McGraw Hill
Illustrative Mathematics®
High School

Cover Credit: alexey_boldin/iStock/Getty Images

mheducation.com/preK-12

Illustrative Math Algebra 1, Geometry, and Algebra 2 are © 2019 Illustrative Mathematics. Modifications © McGraw Hill.

All rights reserved. No part of this publication may be reproduced or distributed in any form or by any means, or stored in a database or retrieval system, without the prior written consent of McGraw Hill, including, but not limited to, network storage or transmission, or broadcast for distance learning.

Send all inquiries to:
McGraw Hill
8787 Orion Place
Columbus, OH 43240

ISBN: 978-1-26-421077-0
MHID: 1-26-421077-9

Illustrative Mathematics, Implementation Guide, High School Math

Printed in the United States of America.

2 3 4 5 6 7 8 9 10 11 12 LMN 28 27 26 25 24 23 22 21 20

'Notice and Wonder' and 'I Notice/I Wonder' are trademarks of the National Council of Teachers of Mathematics, reflecting approaches developed by the Math Forum (http://www.nctm.org/mathforum/), and used here with permission.

Table of Contents

About These Materials..4

Authors and Contributors..5

Design Principles..6

What Is a "Problem-Based" Curriculum?...10

A Typical Lesson...14

How to Use the Materials...17

Instructional Routines...20

Mathematical Modeling Prompts..30

Supporting Diverse Learners..36

    Supporting English Language Learners...36

    Supporting Students with Disabilities..40

Assessments..44

Curriculum Pacing, High School...51

Unit Dependency Chart..52

Required Materials, High School..53

Tables of Contents

    Algebra 1..56

    Algebra 1 Extra Support Materials..60

    Geometry...64

    Algebra 2..68

Correlation to the Common Core State Standards for Mathematics

    Algebra 1..72

    Algebra 1 Extra Support Materials..76

    Geometry...80

    Algebra 2..84

Correlation to the Standards for Mathematical Practice, High School................89

Information for Families...92

About These Materials

These materials were created by Illustrative Mathematics. They were piloted and revised in the 2018–2019 school year.

Course Organization, High School

- Units contain between 10 and 25 lesson plans.
- Each unit has a diagnostic assessment for the beginning of the unit (Check Your Readiness) and an end-of-unit assessment.
- Longer units also have a mid-unit assessment.
- Modeling Prompts are provided to be used throughout the year.

Curriculum Pacing

- The time estimates in these materials refer to instructional time.
- Each lesson plan is designed to fit within a class period that is at least 45 minutes long.
- Some lessons contain optional activities that provide additional scaffolding or practice for teachers to use at their discretion.

Implementation Options

Teachers can access the teacher materials either in print or in a browser. A classroom with a digital projector is recommended. There are two ways students can interact with these materials.

- Students can work solely with printed workbooks or pdfs.
- Alternatively, if all students have access to an appropriate device, students can look at the task statements on that device and write their responses in a notebook or the print companion for the digital materials. It is recommended that if students are to access the materials this way, they keep the notebook carefully organized so that they can go back to their work later.

Materials and Preparation

- Many activities are written in a card sort, matching, or info gap format that requires teachers to provide students with a set of cards or slips of paper that have been photocopied and cut up ahead of time.
- Teachers might stock up on two sizes of resealable plastic bags: sandwich size and gallon size.
- For a given activity, one set of cards can go in each small bag, and then the small bags for one class can be placed in a large bag.
- If these are labeled and stored in an organized manner, it can facilitate preparing ahead of time and re-using card sets between classes.
- Additionally, if possible, it is often helpful to print the slips for different parts of an activity on different color paper. This helps facilitate quickly sorting the cards between classes.

Authors and Contributors

Writing Team

Lauren Baucom
Sandy Berger
Ashli Black, Algebra 2 Lead
Tina Cardone, Geometry Lead
Mimi Cukier
Wendy DenBesten
Nik Doran, Engineering Lead
Angela Harris
Bowen Kerins
Brigitte Lahme
Chuck Larrieu Casias
William McCallum, Shukongojin
Jasmine Moore
Mike Nakamaye
Kate Nowak, Instructional Lead
Carrie Ott
David Petersen, Statistics Lead
Becca Phillips
Max Ray-Riek
Linda Richard
Gabriel Rosenberg
Melissa Schumacher
Benjamin Sinwell
Lizzy Skousen
Yenche Tioanda, Algebra 1 Lead
Kristin Umland, Content Lead

Teacher Professional Learning

Jennifer Wilson
Vanessa Cerrahoglu

Supports for English Language Learners and Students with Disabilities

Erin BeMent
Patricia Gorse
Miyoko Itokazu Bodiford
Sue Jones
Meaghan Krazinski
Liz Ramirez, Lead
Sasha Reese
Moisés Rivera
Erin Smith

Digital Activities Development

Jen Silverman

Copy Editing

Toni Brokaw
Emily Flanagan
Christina Jackyra
Robert Puchalik, Lead
Sue Rice
Rebecca Robinson

Project Management

Sadie Estrella
Olivia Mitchell Russell

Engineering

Eric Connally
Brendan Shean
Jim Whiteman

Image Development

Jonathan Claydon
Tiffany Graves-Davis
Jessica Haase
Cam McLeman, Lead
Matthew Sutter
Siavash Tehrani
Justin Wisby

Image Alt Text

Deb Barnum
Liza Bondurant
Ann Crilley
Mary Cummins
Donna Gustafson
Kia Johnson-Portee, Lead

Content Advisors

Enrique Acosta
Taylor Belcher
Peg Cagle
Amy Callahan
Patrick Callahan
Al Cuoco
Vinci Daro
Joyce Frost
James Malamut
Joe Obrycki
Roxy Peck
Dev Sinha
Sarah Strong
Jade White
Jason Zimba

Design Principles

Developing Conceptual Understanding and Procedural Fluency

1. Each unit begins with a pre-assessment (Check Your Readiness) that helps teachers gauge what students know about both prerequisite and upcoming concepts and skills, so that teachers can gauge where students are and make adjustments accordingly.

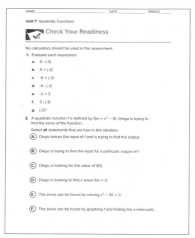

Check Your Readiness, Algebra 1, Unit 7

2. The initial lesson in a unit is designed to activate prior knowledge and provide an easy entry to point to new concepts, so that students at different levels of both mathematical and English language proficiency can engage productively in the work.

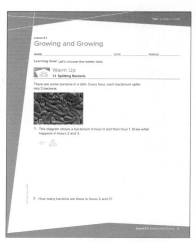

Algebra 1, Lesson 5-1, p.3

3. As the unit progresses, students are systematically introduced to representations, contexts, concepts, language and notation. As their learning progresses, they make connections between different representations and strategies, consolidating their conceptual understanding, and see and understand more efficient methods of solving problems, supporting the shift towards procedural fluency.

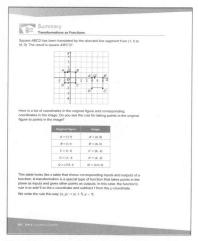

Summary, Geometry, Lesson 6-2, p.150

4. Practice problems, when assigned in a distributed manner, give students ongoing practice, which also supports developing procedural proficiency.

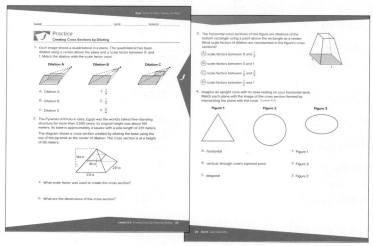

Student Edition Practice, Geometry, Lesson 5-3, pp. 23–24

Applying Mathematics

Students have opportunities to make connections to real-world contexts throughout the materials.

- Frequently, carefully-chosen anchor contexts are used to motivate new mathematical concepts, and students have many opportunities to make connections between contexts and the concepts they are learning.

- Many units include a real-world application lesson at the end.

- In some cases, students spend more time developing mathematical concepts before tackling more complex application problems, and the focus is on mathematical contexts. Additionally, a set of mathematical modeling prompts provide students opportunities to engage in authentic, grade-level appropriate mathematical modeling.

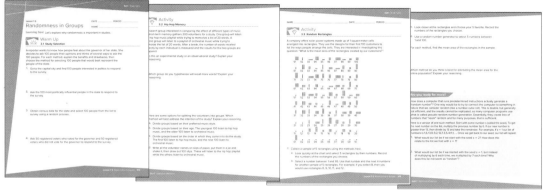

Student Edition, Algebra 2, Lesson 7-3, pp. 413–416

The Five Practices

Selected activities are structured using *Five Practices for Orchestrating Productive Mathematical Discussions* (Smith & Stein, 2011), also described in *Principles to Actions: Ensuring Mathematical Success for All* (NCTM, 2014), and *Intentional Talk: How to Structure and Lead Productive Mathematical Discussions* (Kazemi & Hintz, 2014).

- These activities include a presentation of a task or problem (may be print or other media) where student approaches are anticipated ahead of time.

- Students first engage in independent think-time followed by partner or small-group work on the problem.

- The teacher circulates as students are working and notes groups using different approaches.

- Groups or individuals are selected in a specific, recommended sequence to share their approach with the class, and finally the teacher leads a whole-class discussion to make connections and highlight important ideas.

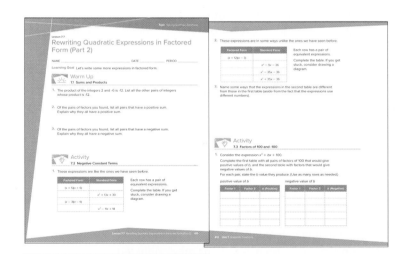
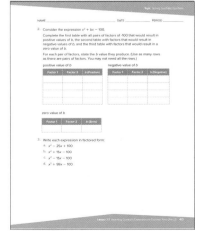

Student Edition, Algebra 1, Lesson 7-7, pp. 411–413

Design Principles 7

Use of Digital Tools

These curriculum materials empower high school teachers and students to become fluent users of widely-accessible mathematical digital tools to produce representations to support their understanding, solve problems, and communicate their reasoning.

Digital tools are included when they are required by the standard being addressed and when they make better learning possible. For example, when a student can use a graphing calculator instead of graphing by hand, use a spreadsheet instead of repeating calculations, or create dynamic geometry drawings instead of making multiple hand-drawn sketches, they can attend to the structure of the mathematics or the meaning of the representation.

Lessons are written with three anticipated levels of digital interaction:

- some activities *require* digital tools
- some activities *suggest* digital tools
- some activities *allow* digital tools
- In a few cases, activities may prohibit digital tools if they interfere with concept development.

In most cases, instead of being given a pre-made applet to explore, students have access to a suite of linked applications, such as:

- graphing tools
- synthetic and analytic geometry tools
- spreadsheets

Students (and teachers) are taught how to use the tools, but not always told when to use them, and student choice in problem-solving approach is valued.

When appropriate, pre-made applets may be included to allow for students to practice many iterations of a skill with error checking, to shorten the amount of time it takes students to create a representation, or to help students see many examples of a relationship in a short amount of time.

Task Purposes

Provide experience with a new context.
Activities that give all students experience with a new context ensure that students are ready to make sense of the concrete before encountering the abstract.

Introduce a new concept and associated language.
Activities that introduce a new concept and associated language build on what students already know and ask them to notice or put words to something new.

Introduce a new representation.
Activities that introduce a new representation often present the new representation of a familiar idea first and ask students to interpret it. Where appropriate, new representations are connected to familiar representations or extended from familiar representations. Students are then given clear instructions on how to create such a representation as a tool for understanding or for solving problems. For subsequent activities and lessons, students are given opportunities to practice using these representations and to choose which representation to use for a particular problem.

Formalize a definition of a term for an idea previously encountered informally.
Activities that formalize a definition take a concept that students have already encountered through examples, and give it a more general definition.

Identify and resolve common mistakes and misconceptions that people make.
Activities that give students a chance to identify and resolve common mistakes and misconceptions usually present some incorrect work and ask students to identify it as such and explain what is incorrect about it. Students deepen their understanding of key mathematical concepts as they analyze and critique the reasoning of others.

Practice using mathematical language.
Activities that provide an opportunity to practice using mathematical language are focused on that as the primary goal rather than having a primarily mathematical learning goal. They are intended to give students a reason to use mathematical language to communicate. These frequently use the Info Gap instructional routine.

Work toward mastery of a concept or procedure.
Activities where students work toward mastery are included for topics where experience shows students often need some additional time to work with the ideas. Often these activities are marked as optional because no new mathematics is covered, so if a teacher were to skip them, no new topics would be missed.

Provide an opportunity to apply mathematics to a modeling or other application problem.
Activities that provide an opportunity to apply mathematics to a modeling or other application problem are most often found toward the end of a unit. Their purpose is to give students experience using mathematics to reason about a problem or situation that one might encounter naturally outside of a mathematics classroom.

A Note about Standards Alignments

There are three kinds of alignments to standards in these materials: building on, addressing, and building towards. Oftentimes a particular standard requires weeks, months, or years to achieve, in many cases building on work in prior grade-levels.

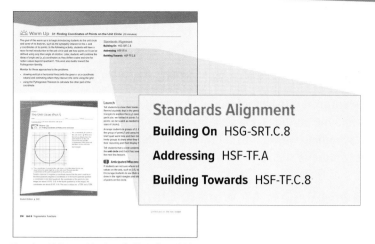

Teacher Edition, Algebra 2, Unit 6, p. 314

Building On	Building Towards	Addressing
When an activity reflects the work of prior grades but is being used to bridge to a grade-level standard, alignments are indicated as "building on."	When an activity is laying the foundation for a grade-level standard but has not yet reached the level of the standard, the alignment is indicated as "building towards."	When a task is focused on the grade-level work, the alignment is indicated as "addressing."

A Note about Mathematical Diagrams

Everything in a mathematical diagram has a mathematical meaning. Students are sense makers looking for connections. The mathematical diagrams provided in activities were designed to include only components with mathematical meaning.

For example, while it is not uncommon to see arrows on the ends of the graph of a function, the arrows add no mathematical meaning to the graph. Arrows are typically used to imply a sense of direction, but a graph of a function is a representation of all the points that make the function true, so there is no direction to imply. It is also possible for students to infer meaning that isn't there, such as assuming the arrows mean the function continues forever in a specific direction. While this idea works for linear functions, it does not work with functions whose graphs curve or are periodic.

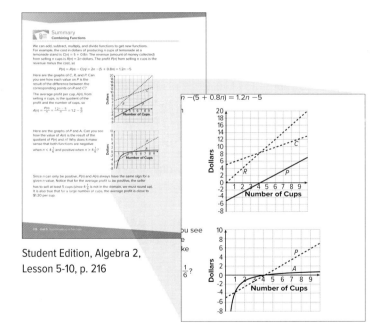

Student Edition, Algebra 2, Lesson 5-10, p. 216

Design Principles

What Is a "Problem-Based" Curriculum?

What Students Should Know and Be Able to Do

Our ultimate purpose is to impact student learning and achievement. First, we define the attitudes and beliefs about mathematics and mathematics learning we want to cultivate in students, and what mathematics students should know and be able to do.

Attitudes and Beliefs We Want to Cultivate

Many people think that mathematical knowledge and skills exclusively belong to "math people." Yet research shows that students who believe that hard work is more important than innate talent learn more mathematics.[1] We want students to believe anyone can do mathematics and that persevering at mathematics will result in understanding and success. In the words of the NRC report *Adding It Up*, we want students to develop a "productive disposition—[the] habitual inclination to see mathematics as sensible, useful, and worthwhile, coupled with a belief in diligence and one's own efficacy."[2]

[1] Uttal, D.H. (1997). Beliefs about genetic influences on mathematics achievement: a cross-cultural comparison. *Genetica*, 99(2-3), 165-172. doi.org/10.1023/A:1018318822120

[2] National Research Council. (2001). *Adding it up: Helping children learn mathematics.* J.Kilpatrick, J. Swafford, and B.Findell (Eds.). Mathematics Learning Study Committee, Center for Education, Division of Behavioral and Social Sciences and Education. Washington, DC: National Academy Press. doi.org/10.17226/9822

Knowledge

Conceptual understanding: Students need to understand the *why* behind the *how* in mathematics. Concepts build on experience with concrete contexts. Students should access these concepts from a number of perspectives in order to see math as more than a set of disconnected procedures.

Procedural fluency: We view procedural fluency as solving problems expected by the standards with speed, accuracy, and flexibility.

Application: Application means applying mathematical or statistical concepts and skills to a novel mathematical or real-world context.

These three aspects of mathematical proficiency are interconnected: procedural fluency is supported by understanding, and deep understanding often requires procedural fluency. In order to be successful in applying mathematics, students must both understand and be able to do the mathematics.

Mathematical Practices

In a mathematics class, students should not just learn *about* mathematics, they should *do* mathematics. This can be defined as engaging in the mathematical practices:

- making sense of problems
- reasoning abstractly and quantitatively
- making arguments and critiquing the reasoning of others
- modeling with mathematics
- making appropriate use of tools
- attending to precision in their use of language
- looking for and making use of structure
- expressing regularity in repeated reasoning

What Teaching and Learning Should Look Like

How teachers should teach depends on what we want students to learn. To understand what teachers need to know and be able to do, we need to understand how students develop the different (but intertwined) strands of mathematical proficiency, and what kind of instructional moves support that development.

Traditional Methods

Teacher tells.
Students listen.

Students practice.
Teacher corrects.

Illustrative Mathematics

1. Teacher ensures students understand the question.

2. Students work individually. Teacher monitors, listens, questions.

3. Students work in groups. Teacher monitors, listens, and asks questions to understand students' thinking.

4. Teacher helps students synthesize their learning.

Principles for Mathematics Teaching and Learning

Active learning is best.

Students learn best and retain what they learn better by solving problems.

Often, mathematics instruction is shaped by the belief that if teachers tell students how to solve problems and then students practice, students will learn how to do mathematics. Decades of research tells us that the traditional model of instruction is flawed.

Traditional instructional methods may get short-term results with procedural skills, but students tend to forget the procedural skills and do not develop problem solving skills, deep conceptual understanding, or a mental framework for how ideas fit together. They also don't develop strategies for tackling non-routine problems, including a propensity for engaging in productive struggle to make sense of problems and persevere in solving them.

Teachers should build on what students know.

New mathematical ideas are built on what students already know about mathematics and the world, and as they learn new ideas, students need to make connections between them.[4] In order to do this, teachers need to understand what knowledge students bring to the classroom and monitor what they do and do not understand as they are learning. Teachers must themselves know how the mathematical ideas connect in order to mediate students' learning.

Good instruction starts with explicit learning goals.

Learning goals must be clear not only to teachers, but also to students, and they must influence the activities in which students participate. Without a clear understanding of what students should be learning, activities in the classroom, implemented haphazardly, have little impact on advancing students' understanding. Strategic negotiation of whole-class discussion on the part of the teacher during an activity synthesis is crucial to making the intended learning goals explicit. Teachers need to have a clear idea of the destination for the day, week, month, and year, and select and sequence instructional activities (or use well-sequenced materials) that will get the class to their destinations. If you are going to a party, you need to know the address and also plan a route to get there; driving around aimlessly will not get you where you need to go.

Different learning goals require different instructional moves.

The kind of instruction that is appropriate at any given time depends on the learning goals of a particular lesson.

Lessons and activities can:

- Introduce students to a new topic of study and invite them to the mathematics
- Study new concepts and procedures deeply
- Integrate and connect representations, concepts, and procedures
- Work towards mastery
- Apply mathematics

Lessons should be designed based on what the intended learning outcomes are. This means that teachers should have a toolbox of instructional moves that they can use as appropriate.

Each and every student should have access to the mathematical work.

With proper structures, accommodations, and supports, all students can learn mathematics. Teachers' instructional toolboxes should include knowledge of and skill in implementing supports for different learners.

[3] Hiebert, J., et. al. (1996). Problem solving as a basis for reform in curriculum and instruction: the case of mathematics. *Educational Researcher* 25(4), 12-21. doi.org/10.3102/0013189X025004012

[4] National Research Council. (2001). *Adding it up: Helping children learn mathematics*. J.Kilpatrick, J. Swafford, and B.Findell (Eds.). Mathematics Learning Study Committee, Center for Education, Division of Behavioral and Social Sciences and Education. Washington, DC: National Academy Press. doi.org/10.17226/9822

The Student's Role	The Teacher's Role
In order to learn mathematics, students should spend time in math class *doing mathematics*. *"Students learn mathematics as a result of solving problems. Mathematical ideas are the outcomes of the problem-solving experience rather than the elements that must be taught before problem solving."*[3] Students should take an active role, both individually and in groups, to see what they can figure out before having things explained to them or being told what to do.	Teachers play a critical role in mediating student learning, but that role looks different than simply showing, telling, and correcting. The teacher's role is … 1. to ensure students understand the context and what is being asked, 2. ask questions to advance students' thinking in productive ways, 3. help students share their work and understand others' work through orchestrating productive discussions, and 4. synthesize the learning with students at the end of activities and lessons.

Critical Practices

Intentional planning

Because different learning goals require different instructional moves, teachers need to be able to plan their instruction appropriately. While a high-quality curriculum does reduce the burden for teachers to create or curate lessons and tasks, it does not reduce the need to spend deliberate time planning lessons and tasks.

Instead, teachers' planning time can shift to high-leverage practices (practices that teachers without a high-quality curriculum often report wishing they had more time for):

- reading and understanding the high-quality curriculum materials
- identifying connections to prior and upcoming work
- diagnosing students' readiness to do the work
- leveraging instructional routines to address different student needs and differentiate instruction
- anticipating student responses that will be important to move the learning forward
- planning questions and prompts that will help students attend to, make sense of, and learn from each other's work
- planning supports and extensions to give as many students as possible access to the main mathematical goals
- figuring out timing, pacing, and opportunities for practice
- preparing necessary supplies
- and the never-ending task of giving feedback on student work.

Establishing norms

Norms around doing math together and sharing understandings play an important role in the success of a problem-based curriculum. For example, students must feel safe taking risks, listen to each other, disagree respectfully, and honor equal air time when working together in groups. Establishing norms helps teachers cultivate a community of learners where making thinking visible is both expected and valued.

Building a shared understanding of a small set of instructional routines

Instructional routines allow the students and teacher to become familiar with the classroom choreography and what they are expected to do. This means that they can pay less attention to what they are supposed to do and more attention to the mathematics to be learned. Routines can provide a structure that helps strengthen students' skills in communicating their mathematical ideas.

Using high quality curriculum

A growing body of evidence suggests that using a high-quality, coherent curriculum can have a significant impact on student learning.[5] Creating a coherent, effective instructional sequence from the ground up takes significant time, effort, and expertise. Teaching is already a full-time job, and adding curriculum development on top of that means teachers are overloaded or shortchanging their students.

Ongoing formative assessment

Teachers should know what mathematics their students come into the classroom already understanding, and use that information to plan their lessons. As students work on problems, teachers should ask questions to better understand students' thinking, and use expected student responses and potential misconceptions to build on students' mathematical understanding during the lesson. Teachers should monitor what their students have learned at the end of the lesson and use this information to provide feedback and plan further instruction.

[5] Steiner, D. (2017). Curriculum research: What we know and where we need to go. *Standards Work*. Retrieved from https://standardswork.org/wp-content/uploads/2017/03/sw-curriculum-research-report-fnl.pdf

A Typical Lesson

A typical lesson has four phases.

1. Warm Up
2. One or More Instructional Activities
3. Lesson Synthesis
4. Cool Down

Warm Up (5-10 minutes)

The first event in every lesson is a Warm Up.

A Warm Up either:

1. helps students get ready for the day's lesson, or
2. gives students an opportunity to strengthen their number sense or procedural fluency.

1. Get Ready for Today's Lesson

A warm up that helps students get ready for today's lesson might serve to remind them of a context they have seen before, get them thinking about where the previous lesson left off, or preview a calculation that will happen in the lesson so that the calculation doesn't get in the way of learning new mathematics.

2. Strengthen Number Sense or Procedural Fluency

A warm up that is meant to strengthen number sense or procedural fluency asks students to do mental arithmetic or reason numerically or algebraically. It gives them a chance to make deeper connections or become more flexible in their thinking.

What if the Warm Ups are taking longer than anticipated?

Once students and teachers become used to the routine, warm ups should take 5–10 minutes. If warm ups frequently take much longer than that, the teacher should work on concrete moves to more efficiently accomplish the goal of the warm up.

Frequently Used Instructional Routines

Instructional routines frequently used in Warm Ups are Notice and Wonder, and Which One Doesn't Belong. In addition to the mathematical purposes, these routines serve the additional purpose of strengthening students' skills in listening and speaking about mathematics.

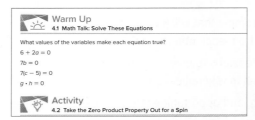

Math Talk, Algebra 1, Lesson 7-4, p. 385

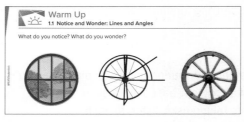

Notice and Wonder, Geometry, Lesson 7-1, p. 275

Classroom Tip

At the beginning of the year, consider establishing a small, discreet hand signal students can display to indicate they have an answer they can support with reasoning. This signal could be a thumbs up, or students could show the number of fingers that indicate the number of responses they have for the problem. This is a quick way to see if students have had enough time to think about the problem and keeps them from being distracted or rushed by classmates' raised hands.

Classroom Activities (Pacing Varies)

- After the Warm Up, lessons consist of a sequence of one to three classroom activities.
- The activities are the heart of the mathematical experience and make up the majority of the time spent in class.
- An activity can serve one or more of many purposes.
  - Provide experience with a new context.
  - Introduce a new concept and associated language.
  - Introduce a new representation.
  - Formalize a definition of a term for an idea previously encountered informally.
  - Identify and resolve common mistakes and misconceptions that people make.
  - Practice using mathematical language.
  - Work toward mastery of a concept or procedure.
  - Provide an opportunity to apply mathematics to a modeling or other application problem.
- The purpose of each activity is described in its Activity Narrative. Read more about how activities serve these different purposes in the section on Design Principles.

Activity 8.2, *Under One Condition*, Geometry, Lesson 8-8, p. 458

A Note about Optional Activities

A relatively small number of activities throughout the course have been marked "optional." Some common reasons an activity might be optional include the following.

Reason for Optional Activity	Guidance for Use
The activity addresses a concept or skill that is **below grade level**, but we know that it is common for students to need a chance to focus on it before encountering grade-level material.	If the pre-unit diagnostic assessment (Check Your Readiness) indicates that students don't need this review, an activity like this can be safely skipped.
The activity addresses a concept or skill that goes **beyond the requirements of a standard**.	The activity is nice to do if there is time, but students won't miss anything important if the activity is skipped.
The activity provides an opportunity for **additional practice** on a concept or skill that we know many students (but not necessarily all students) need.	Teachers should use their judgment about whether class time is needed for such an activity.

A Typical Lesson 15

Lesson Synthesis (5-10 min)

- After the activities for the day are done, students should take time to synthesize what they have learned.
- This portion of class should take 5–10 minutes before students start working on the Cool Down.
- Each lesson includes a Lesson Synthesis section that assists the teacher with ways to help students incorporate new insights gained during the activities into their big-picture understanding.
- Teachers can use this time in any number of ways, including...
  - posing questions verbally and calling on volunteers to respond,
  - asking students to respond to prompts in a written journal,
  - asking students to add on to a graphic organizer or concept map, or
  - adding a new component to a persistent display like a word wall.

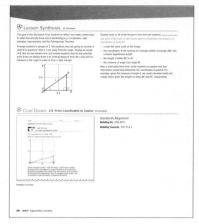

Lesson Synthesis
Algebra 2, Lesson 6-2, p. 310

Cool Down (5 min)

- Each lesson includes a Cool Down task to be given to students at the end of the lesson.
- Students are meant to work on the Cool Down for about 5 minutes independently and turn it in.
- The Cool Down serves as a brief formative assessment to determine whether students understood the lesson.
- Students' responses to the Cool Down can be used to make adjustments to further instruction.

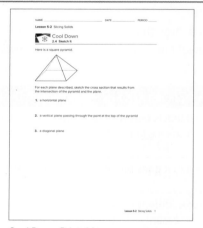

Cool Down Printable
Geometry, Lesson 5-2

16 A Typical Lesson

How to Use the Materials

Each Lesson and Unit Tells a Story

- This story each course is told in seven or eight units.
- Each unit has a narrative that describes the mathematical work that will unfold in that unit.
- Each lesson in the unit also has a narrative.

Lesson Narratives explain:

- A description of the mathematical content of the lesson and its place in the learning sequence.
- The meaning of any new terms introduced in the lesson.
- How the mathematical practices come into play, as appropriate.

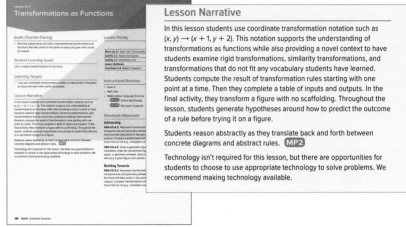

Teacher Edition, Geometry,
Lesson 6-2, p. 188

Activities within lessons also have a narrative, which explain:

- The mathematical purpose of the activity and its place in the learning sequence.
- What students are doing during the activity.
- What teacher needs to look for while students are working on an activity to orchestrate an effective synthesis.
- Connections to the mathematical practices when appropriate.

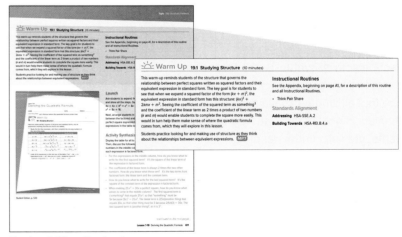

Teacher Edition, Algebra 1,
Lesson 7-19, p. 611

Launch – Work – Synthesize

Each classroom activity has three phases.

Launch

During the launch, the teacher makes sure that students understand the context (if there is one) and *what the problem is asking them to do*. This is not the same as making sure the students know *how to do the problem*—part of the work that students should be doing for themselves is figuring out how to solve the problem.

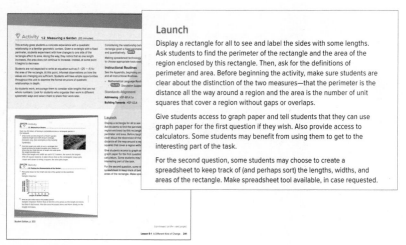

Teacher Edition, Algebra 1,
Lesson 6-1, p. 241

Student Work Time

The launch for an activity frequently includes suggestions for grouping students. This gives students the opportunity to work individually, with a partner, or in small groups.

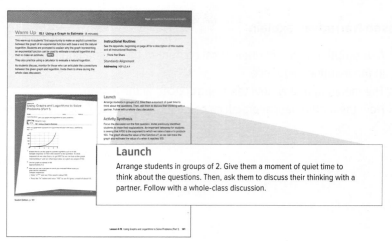

Teacher Edition, Algebra 2,
Lesson 4-15, p. 141

Activity Synthesis

During the activity synthesis, the teacher orchestrates some time for students to synthesize what they have learned. This time is used to ensure that all students have an opportunity to understand the mathematical goal of the activity and situate the new learning within students' previous understanding.

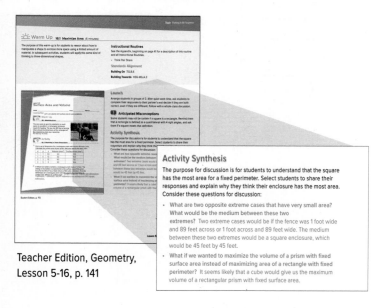

Teacher Edition, Geometry,
Lesson 5-16, p. 141

Practice Problems

Each lesson includes an associated set of practice problems. Teachers may decide to assign practice problems for homework or for extra practice in class. They may decide to collect and score it or to provide students with answers ahead of time for self-assessment. It is up to teachers to decide which problems to assign (including assigning none at all).

The practice problems associated with each lesson include a few questions about the contents of that lesson, plus additional problems that review material from earlier in the unit and previous units. Distinguished practice (revisiting that same content over time) is more effective than massed practice (a large amount of practice on one topic, but all at once).

Are you ready for more?

Select classroom activities include an opportunity for differentiation for students ready for more of a challenge. We think of them as the "mathematical dessert" to follow the "mathematical entrée" of a classroom activity.

Student Edition, Are you ready for more?, Geometry, Lesson 7-9, p. 341

Every extension problem is made available to all students with the heading "Are you ready for more?"

- These problems go deeper into grade-level mathematics and often make connections between the topic at hand and other concepts.
- Some of these problems extend the work of the associated activity, but some of them involve work from prior grades, prior units in the course, or reflect work that is related to the K-12 curriculum but a type of problem not required by the standards.
- They are not routine or procedural, and they are not just "the same thing again but with harder numbers."

- They are intended to be used on an opt-in basis by students if they finish the main class activity early or want to do more mathematics on their own.
- It is not expected that an entire class engages in *Are you ready for more?* problems, and it is not expected that any student works on all of them.
- *Are you ready for more?* problems may also be good fodder for a Problem of the Week or similar structure.

Instructional Routines

The kind of instruction appropriate in any particular lesson depends on the learning goals of that lesson. Some lessons may be devoted to developing a concept, others to mastering a procedural skill, yet others to applying mathematics to a real-world problem. These aspects of mathematical proficiency are interwoven. These materials include a small set of activity structures and reference a small, high-leverage set of teacher moves that become more and more familiar to teachers and students as the year progresses.

The first instance of each routine in a course includes more detailed guidance for how to successfully conduct the routine. Subsequent instances include more abbreviated guidance, so as not to unnecessarily inflate the word count of the teacher guide.

Digital Routines indicate required or suggested applications of technology, appearing repeatedly throughout the curriculum. Activities using the routines are flagged for the teacher, which is helpful for lesson planning and for focusing the work of professional development.

Some of the instructional routines, known as Mathematical Language Routines (MLR), were developed by the Stanford University UL/SCALE team. The purpose of each MLR is described on the next several pages, but you can read more about supports for students with emerging English language proficiency in the Supporting English Language Learners section.

Instructional Routines
Analyze It
Anticipate, Monitor, Select, Sequence, Connect
Aspects of Mathematical Modeling
Card Sort
Construct It
Draw It
Estimation
Extend It
Fit it
Graph It
Math Talk
Mathematical Language Routines
MLR1 Stronger and Clearer Each Time
MLR2 Collect and Display
MLR3 Clarify, Critique, and Correct
MLR4 Information Gap Cards
MLR5 Co-Craft Questions
MLR6 Three Reads
MLR7 Compare and Connect
MLR8 Discussion Supports
Notice and Wonder
Poll the Class
Take Turns
Think Pair Share
Which One Doesn't Belong?

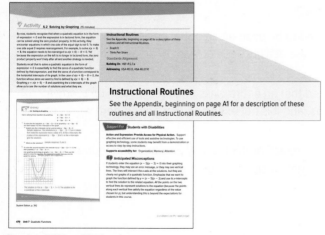

Teacher Edition, Algebra 1,
Lesson 7-5, p. 470

Descriptions of the Instructional Routine, including the Mathematical Language Routines, are located on the next several pages. Descriptions are also located in the Appendix of the Teacher Edition, beginning on page A1.

Analyze It

Analyze It indicates activities where students have an opportunity to use statistical tools to calculate and display numeric statistics and produce visual representations of one- and two-variable data sets.

Anticipate, Monitor, Select, Sequence, Connect

- **What** These are the *5 Practices for Orchestrating Productive Mathematical Discussions* (Smith and Stein, 2011). In this curriculum, much of the work of anticipating, sequencing, and connecting is handled by the materials in the activity narrative, launch, and synthesis sections. Teachers need to prepare for and conduct whole-class discussions.
- **Where** Many classroom activities lend themselves to this structure.
- **Why** In a problem-based curriculum, many activities can be described as "do math and talk about it," but the 5 Practices lend more structure to these activities so that they more reliably result in students making connections and learning new mathematics.

Aspects of Mathematical Modeling

- **What** In activities tagged with this routine, students engage in scaled-back modeling scenarios, for which students only need to engage in a part of a full modeling cycle. For example, they may be selecting quantities of interest in a situation or choosing a model from a list.
- **Why** Mathematical modeling is often new territory for both students and teachers. Opportunities to develop discrete skills in the supported environment of a classroom lesson make success more likely when students engage in more open-ended modeling.

Card Sort

- **What** A Card Sort uses cards or slips of paper that can be manipulated and moved around (or the same functionality enacted with a computer interface). It can be done individually or in groups of 2–4. Students put things into categories or groups based on shared characteristics or connections. This routine can be combined with Take Turns, such that each time a student sorts a card into a category or makes a match, they are expected to explain the rationale while the group listens for understanding. The first few times students engage in these activities, the teacher should demonstrate how the activity is expected to go. Once students are familiar with these structures, less set-up will be necessary. While students are working, the teacher can ask students to restate their question more clearly or paraphrase what their partner said.
- **Where** Classroom Activities
- **Why** A Card Sort provides opportunities to attend to mathematical connections using representations that are already created, instead of expending time and effort generating representations. It gives students opportunities to analyze representations, statements, and structures closely and make connections (MP2, MP7).

Construct It

Construct It indicates activities where students have an opportunity to use digital tools that are equivalent to those of classical geometry to construct figures.

Draw It

Draw It indicates an activity where students have an opportunity to use digital geometry tools to produce a visual representation of a problem or scenario.

Estimation

In the Estimation routine, an image is displayed for all to see. Students silently think of a number they are sure is too low, a number they are sure is too high, and a number that is about right, and write these down. Then, they write a short explanation for the reasoning behind their estimate. A few students share their estimates, and then the actual amount is revealed. Finally, the class has an opportunity to reflect on how accurate their estimates were.

Purpose To practice the skill of estimating a reasonable answer based on experience and known information, and also help students develop a deeper understanding of the meaning of standard units of measure.

- It gives students a low-stakes opportunity to share a mathematical claim and the thinking behind it (MP3).
- Asking yourself "Does this make sense?" is a component of making sense of problems (MP1), and making an estimate or a range of reasonable answers with incomplete information is a part of modeling with mathematics (MP4).
- Estimation warm-ups are intended to improve students' number sense over time, and aren't necessarily connected to the work in the lessons in which they appear.

The Estimation instructional routine was inspired by the work of Andrew Stadel at Estimation180.com.

Extend It

Extend It indicates activities where students have an opportunity to use a spreadsheet to produce a sequence of numbers to see patterns and make predictions.

Fit It

Fit It indicates activities where students have an opportunity to use a table of points to produce a graph to see patterns and make predictions. Also when appropriate, find a function that best fits the data.

Graph It

Graph It indicates activities where students have an opportunity to use graphing technology to visualize a graph representing one or more functions with known parameters and use the tool to find features like intersection points, intercepts, and maximums or minimums. Additionally, they may use sliders for exploring the effect of changing parameters.

Math Talk

- **What** In these warm-ups, one problem is displayed at a time. Students are given a few moments to quietly think and give a signal when they have an answer and a strategy. The teacher selects students to share different strategies for each problem, asking, "Who thought about it a different way?" Their explanations are recorded for all to see. Students might be pressed to provide more details about why they decided to approach a problem a certain way. It may not be possible to share every possible strategy in the given time—the teacher may only gather two or three distinctive strategies per problem. Problems are purposefully chosen to elicit different approaches, often in a way that builds from one problem to the next.

- **Why** Math Talks build fluency by encouraging students to think about the numbers, shapes, or algebraic expressions and rely on what they know about structure, patterns, and properties of operations to mentally solve a problem. While participating in these activities, students need to be precise in their word choice and use of language (MP6). Additionally a Math Talk often provides opportunities to notice and make use of structure (MP7).

Mathematical Language Routines (MLR)

MLR1 Stronger and Clearer Each Time *Adapted from Zwiers (2014)*

Purpose To provide a structured and interactive opportunity for students to revise and refine both their ideas and their verbal and written output (Zwiers, 2014). This routine also provides a purpose for student conversation through the use of a discussion-worthy and iteration-worthy prompt.

- The main idea is to have students think and write individually about a question, use a structured pairing strategy to have multiple opportunities to refine and clarify their response through conversation, and then finally revise their original written response.
- Subsequent conversations and second drafts should naturally show evidence of incorporating or addressing new ideas and language. They should also show evidence of refinement in precision, communication, expression, examples, and reasoning about mathematical concepts.

How it happens

Prompt: This routine begins by providing a thought-provoking question or prompt. The prompt should guide students to think about a concept or big idea connected to the content goal of the lesson, and should be answerable in a format that is connected with the activity's primary disciplinary language function.

Response - First Draft: Students draft an initial response to the prompt by writing or drawing their initial thoughts in a first draft. Responses should attempt to align with the activity's primary language function. It is not necessary that students finish this draft before moving to the structured pair meetings step. However, students should be encouraged to write or draw something before meeting with a partner. This encouragement can come over time as class culture is developed, strategies and supports for getting started are shared, and students become more comfortable with the low stakes of this routine. (2–3 min)

Structured Pair Meetings: Next, use a structured pairing strategy to facilitate students having 2–3 meetings with different partners. Each meeting gives each partner an opportunity to be the speaker and an opportunity to be the listener. As the speaker, each student shares their ideas (without looking at their first draft, when possible). As a listener, each student should (a) ask questions for clarity and reasoning, (b) press for details and examples, and (c) give feedback that is relevant for the language goal. (1–2 min each meeting)

Response - Second Draft: Finally, after meeting with 2–3 different partners, students write a second draft. This draft should naturally reflect borrowed ideas from partners, as well as refinement of initial ideas through repeated communication with partners. This second draft will be stronger (with more or better evidence of mathematical content understanding) and clearer (more precision, organization, and features of disciplinary language function). After students are finished, their first and second drafts can be compared. (2–3 min)

MLR2 Collect and Display

Purpose To capture a variety of students' oral words and phrases into a stable, collective reference.

- The intent of this routine is to stabilize the varied and fleeting language in use during mathematical work, in order for students' own output to become a reference in developing mathematical language.
- The teacher listens for, and scribes, the language students use during partner, small group, or whole class discussions using written words, diagrams and pictures.
- This collected output can be organized, revoiced, or explicitly connected to other language in a display that all students can refer to, build on, or make connections with during future discussion or writing.
- Throughout the course of a unit (and beyond), teachers can reference the displayed language as a model, update and revise the display as student language changes, and make bridges between prior student language and new disciplinary language (Dieckman, 2017).
- This routine provides feedback for students in a way that supports sense-making while simultaneously increasing meta-awareness of language.

How it happens

Collect: During this routine, circulate and listen to student talk during paired, group, or as a whole-class discussion. Jot down the words, phrases, drawings, or writing students use. Capture a variety of uses of language that can be connected to the lesson content goals, as well as the relevant disciplinary language function(s). Collection can happen digitally, or with a clipboard, or directly onto poster paper; capturing on a whiteboard is not recommended due to risk of erasure.

Display: Display the language collected visually for the whole class to use as a reference during further discussions throughout the lesson and unit. Encourage students to suggest revisions, updates, and connections be added to the display as they develop—over time—both new mathematical ideas and new ways of communicating ideas. The display provides an opportunity to showcase connections between student ideas and new vocabulary. It also provides opportunity to highlight examples of students using disciplinary language functions, beyond just vocabulary words.

MLR3 Clarify, Critique, and Correct

Purpose To give students a piece of mathematical writing that is not their own to analyze, reflect on, and develop.

- The intent is to prompt student reflection with an incorrect, incomplete, or ambiguous written mathematical statement, and for students to improve upon the written work by correcting errors and clarifying meaning.
- Teachers can demonstrate how to effectively and respectfully critique the work of others with meta-think-alouds and pressing for details when necessary.
- This routine fortifies output and engages students in meta-awareness. More than just error analysis, this routine purposefully engages students in considering both the author's mathematical thinking as well as the features of their communication.

How it happens

Original Statement: Create or curate a written mathematical statement that intentionally includes conceptual (or common) errors in mathematical thinking as well as ambiguities in language. The mathematical errors should be driven by the content goals of the lesson and the language ambiguities should be driven by common or typical challenges with the relevant disciplinary language function. This mathematical text is read by the students and used as the draft, or "original statement," that students improve. (1–2 min)

Discussion with Partner: Next, students discuss the original statement in pairs. The teacher provides guiding questions for this discussion such as, "What do you think the author means?," "Is anything unclear?," or "Are there any reasoning errors?" In addition to these general guiding questions, 1–2 questions can be added that specifically address the content goals and disciplinary language function relevant to the activity. (2–3 min)

Improved Statement: Students individually revise the original statement, drawing on the conversations with their partners, to create an "improved statement." In addition to resolving any mathematical errors or misconceptions, clarifying ambiguous language, other requirements can be added as parameters for the improved response. These specific requirements should be aligned with the content goals and disciplinary language function of the activity. (3–5 min)

MLR4 Information Gap *Adapted from Zwiers 2004*

Purpose To create a need for students to communicate (Gibbons, 2002). This routine allows teachers to facilitate meaningful interactions by positioning some students as holders of information that is needed by other students.

- The information is needed to accomplish a goal, such as solving a problem or winning a game. With an information gap, students need to orally (or visually) share ideas and information in order to bridge a gap and accomplish something that they could not have done alone.
- Teachers should demonstrate how to ask for and share information, how to justify a request for information, and how to clarify and elaborate on information. This routine cultivates conversation.

How it happens

Problem/Data Cards: Students are paired into Partner A and Partner B. Partner A is given a card with a problem that must be solved, and Partner B has the information needed to solve it on a "data card." Data cards can also contain diagrams, tables, graphs, etc. Neither partner should read nor show their cards to their partners. Partner A determines what information they need, and prepares to ask Partner B for that specific information. Partner B should not share the information unless Partner A specifically asks for it and justifies the need for the information.

Because partners don't have the same information, Partner A must work to produce clear and specific requests, and Partner B must work to understand more about the problem through Partner A's requests and justifications.

Bridging the Gap

- Partner B asks "What specific information do you need?" Partner A asks for specific information from Partner B.
- Before sharing the requested information, Partner B asks Partner A for a justification: "Why do you need that information?"
- Partner A explains how they plan to use the information.
- Partner B asks clarifying questions as needed, and then provides the information.
- These four steps are repeated until Partner A is satisfied that they have information they need to solve the problem.

Solving the Problem

- Partner A shares the problem card with Partner B. Partner B does not share the data card.
- Both students solve the problem independently, then discuss their strategies. Partner B can share the data card after discussing their independent strategies.

MLR5 Co-Craft Questions

Purpose To allow students to get inside of a context before feeling pressure to produce answers, to create space for students to produce the language of mathematical questions themselves, and to provide opportunities for students to analyze how different mathematical forms and symbols can represent different situations.

- Through this routine, students are able to use conversation skills to generate, choose (argue for the best one), and improve questions and situations as well as develop meta-awareness of the language used in mathematical questions and problems.

How it happens

Hook: Begin by presenting students with a hook—a context or a stem for a problem, with or without values included. The hook can also be a picture, video, or list of interesting facts.

Students Write Questions: Next, students write down possible mathematical questions that might be asked about the situation. These should be questions that they think are answerable by doing math and could be questions about the situation, information that might be missing, and even about assumptions that they think are important. (1–2 minutes)

Students Compare Questions: Students compare the questions they generated with a partner (1–2 minutes) before sharing questions with the whole class. Demonstrate (or ask students to demonstrate) identifying specific questions that are aligned to the content goals of the lesson as well as the disciplinary language function. If there are no clear examples, teachers can demonstrate adapting a question or ask students to adapt questions to align with specific content or function goals. (2–3 minutes)

Actual Question(s) Revealed/Identified: Finally, the actual questions students are expected to work on are revealed or selected from the list that students generated.

MLR6 Three Reads

Purpose To ensure that students know what they are being asked to do, create opportunities for students to reflect on the ways mathematical questions are presented, and equip students with tools used to actively make sense of mathematical situations and information (Kelemanik, Lucenta, & Creighton, 2016).

- This routine supports reading comprehension, sense-making, and meta-awareness of mathematical language.
- It also supports negotiating information in a text with a partner through mathematical conversation.

How it happens

In this routine, students are supported in reading a mathematical text, situation, or word problem three times, each with a particular focus. The intended question or main prompt is intentionally withheld until the third read so that students can concentrate on making sense of what is happening in the text before rushing to a solution or method.

Read #1: Shared Reading (one person reads aloud while everyone else reads with them) The first read focuses on the situation, context, or main idea of the text. After a shared reading, ask students "what is this situation about?" This is the time to identify and resolve any challenges with any non-mathematical vocabulary. (1 minute)

Read #2: Individual, Pairs, or Shared Reading After the second read, students list any quantities that can be counted or measured. Students are encouraged not to focus on specific values. Instead they focus on naming what is countable or measurable in the situation. It is not necessary to discuss the relevance of the quantities, just to be specific about them (examples: "number of people in her family" rather than "people," "number of markers after" instead of "markers"). Some of the quantities will be explicit (example: 32 apples) while others are implicit (example: the time it takes to brush one tooth). Record the quantities as a reference to use when solving the problem after the third read. (3–5 minutes)

Read #3: Individual, Pairs, or Shared Reading During the third read, the final question or prompt is revealed. Students discuss possible solution strategies, referencing the relevant quantities recorded after the second read. It may be helpful for students to create diagrams to represent the relationships among quantities identified in the second read, or to represent the situation with a picture (Asturias, 2014). (1–2 minutes)

MLR7 Compare and Connect

Purpose To foster students' meta-awareness as they identify, compare, and contrast different mathematical approaches and representations.

- This routine leverages the powerful mix of disciplinary representations available in mathematics as a resource for language development.
- In this routine, students make sense of mathematical strategies other than their own by relating and connecting other approaches to their own.
- Students should be prompted to reflect on, and linguistically respond to, these comparisons (for example, exploring why or when one might do or say something a certain way, identifying and explaining correspondences between different mathematical representations or methods, or wondering how a certain concept compares or connects to other concepts).
- Be sure to demonstrate asking questions that students can ask each other, rather than asking questions to "test" understanding.
- Use think alouds to demonstrate the trial and error, or fits and starts of sense-making (similar to the way teachers think aloud to demonstrate reading comprehension).
- This routine supports metacognition and metalinguistic awareness, and also supports constructive conversations.

How it happens

Students Prepare Displays of their Work: Students are given a problem that can be approached and solved using multiple strategies, or a situation that can be modeled using multiple representations. Students are assigned the job of preparing a visual display of how they made sense of the problem and why their solution makes sense. Variation is encouraged and supported among the representations that different students use to show what makes sense.

Compare: Students investigate each others' work by taking a tour of the visual displays. Tours can be self-guided, a "travellers and tellers" format, or the teacher can act as "docent" by providing questions for students to ask of each other, pointing out important mathematical features, and facilitating comparisons. Comparisons focus on the typical structures, purposes, and affordances of the different approaches or representations: what worked well in this or that approach, or what is especially clear in this or that representation. During this discussion, listen for and amplify any comments about what might make this or that approach or representation more complete or easy to understand.

Connect: The discussion then turns to identifying correspondences between different representations. Students are prompted to find correspondences in how specific mathematical relationships, operations, quantities, or values appear in each approach or representation. Guide students to refer to each other's thinking by asking them to make connections between specific features of expressions, tables, graphs, diagrams, words, and other representations of the same mathematical situation. During the discussion, amplify language students use to communicate about mathematical features that are important for solving the problem or modeling the situation. Call attention to the similarities and differences between the ways those features appear.

MLR8 Discussion Supports

To support rich discussions about mathematical ideas, representations, contexts, and strategies (Chapin, O'Connor, & Anderson, 2009).

Rather than another structured format, the examples provided in this routine are instructional moves that can be combined and used together with any of the other routines.

- They include multimodal strategies for helping students make sense of complex language, ideas, and classroom communication.
- The examples can be used to invite and incentivize more student participation, conversation, and meta-awareness of language.
- Eventually, as teachers continue to demonstrate, students should begin using these strategies themselves to prompt each other to engage more deeply in discussions.

How it happens

Unlike the other routines, this support is a collection of strategies and moves that can be combined and used to support discussion during almost any activity.

Examples of possible strategies:

- Revoice student ideas to demonstrate mathematical language use by restating a statement as a question in order to clarify, apply appropriate language, and involve more students.
- Press for details in students' explanations by requesting for students to challenge an idea, elaborate on an idea, or give an example.
- Show central concepts multi-modally by using different types of sensory inputs: acting out scenarios or inviting students to do so, showing videos or images, using gesture, and talking about the context of what is happening.
- Practice phrases or words through choral response.
- Think aloud by talking through thinking about a mathematical concept while solving a related problem or doing a task.
- Demonstrate uses of disciplinary language functions such as detailing steps, describing and justifying reasoning, and questioning strategies.

- Give students time to make sure that everyone in the group can explain or justify each step or part of the problem. Then make sure to vary who is called on to represent the work of the group so students get accustomed to preparing each other to fill that role.
- Prompt students to think about different possible audiences for the statement, and about the level of specificity or formality needed for for a classmate vs. a mathematician, for example. [Convince Yourself, Convince a Friend, Convince a Skeptic (Mason, Burton, & Stacey, 2010)]

Sentence Frames

Sentence frames can support student language production by providing a structure to communicate about a topic. Helpful sentence frames are open-ended, so as to amplify language production, not constrain it. The table shows examples of generic sentence frames that can support common disciplinary language functions across a variety of content topics. Some of the lessons in these materials include suggestions of additional sentence frames that could support the specific content and language functions of that lesson.

Language Function	Sample Sentence Frames	Language Function	Sample Sentence Frames
describe	• It looks like… • I notice that… • I wonder if… • Let's try… • A quantity that varies is _____. • What do you notice? • What other details are important?	critique	• That could/couldn't be true because… • This method works/doesn't work because… • We can agree that… • _____'s idea reminds me of… • Another strategy would be _____ because… • Is there another way to say/do…?
explain	• First, I _____ because… • Then/Next, I… • I noticed _____ so I… • I tried _____ and what happened was… • How did you get…? • What else could we do?	compare and contrast	• Both _____ and _____ are alike because… • _____ and _____ are different because… • One thing that is the same is… • One thing that is different is… • How are _____ and _____ different? • What do _____ and _____ have in common?
justify	• I know _____ because… • I predict _____ because… • If _____ then _____ because… • Why did you…? • How do you know…? • Can you give an example?	represent	• _____ represents _____. • _____ stands for _____. • _____ corresponds to _____. • Another way to show _____ is… • How else could we show this?
generalize	• _____ reminds me of _____ because… • _____ will always _____ because… • _____ will never _____ because… • Is it always true that…? • Is _____ a special case?	interpret	• We are trying to… • We will need to know… • We already know… • It looks like _____ represents… • Another way to look at it is… • What does this part of _____ mean? • Where does _____ show…?

References

Aguirre, J. M. & Bunch, G. C. (2012). What's language got to do with it?: Identifying language demands in mathematics instruction for English language learners. In S. Celedón-Pattichis & N. Ramirez (Eds.), *Beyond good teaching: Advancing mathematics education for ELLs*. (pp. 183-194). Reston, VA: National Council of Teachers of Mathematics.

Chapin, S., O'Connor, C., & Anderson, N. (2009). *Classroom discussions: Using math talk to help students learn, grades K-6* (second edition). Sausalito, CA: Math Solutions Publications.

Gibbons, P. (2002). *Scaffolding language, scaffolding learning: Teaching second language learners in the mainstream classroom*. Portsmouth, NH: Heinemann.

Kelemanik, G, Lucenta, A & Creighton, S.J. (2016). *Routines for reasoning: Fostering the mathematical practices in all students*. Portsmouth, NH: Heinemann.

Zwiers, J. (2011). *Academic conversations: Classroom talk that fosters critical thinking and content understandings*. Portland, ME: Stenhouse

Zwiers, J. (2014). *Building academic language: Meeting Common Core Standards across disciplines, grades 5–12* (2nd ed.). San Francisco, CA: Jossey-Bass.

Zwiers, J., Dieckmann, J., Rutherford-Quach, S., Daro, V., Skarin, R., Weiss, S., & Malamut, J. (2017). *Principles for the design of mathematics curricula: Promoting language and content development*. Retrieved from Stanford University, UL/SCALE website: http://ell.stanford.edu/content/mathematics-resources-additional-resources

Notice and Wonder

- **What** This routine can appear as a warm up or in the launch or synthesis of a classroom activity. Students are shown some media or a mathematical representation. The prompt to students is "What do you notice? What do you wonder?" Students are given a few minutes to think of things they notice and things they wonder and share them with a partner. Then, the teacher asks several students to share things they noticed and things they wondered; these are recorded by the teacher for all to see. Sometimes, the teacher steers the conversation to wondering about something mathematical that the class is about to focus on.

- **Where** Appears frequently in Warm Ups but also appears in launches to classroom activities.

- **Why** The purpose is to make a mathematical task accessible to all students with these two approachable questions. By thinking about them and responding, students gain entry into the context and might get their curiosity piqued. Taking steps to become familiar with a context and the mathematics that might be involved is making sense of problems (MP1).

Note: *Notice and Wonder* and *I Notice/I Wonder* are trademarks of NCTM and the Math Forum and used in these materials with permission.

Poll the Class

- **What** This routine is used to register an initial response or an estimate, most often in activity launches or to kick off a discussion. It can also be used when data needs to be collected from each student in class, for example, "What is the length of your ear in centimeters?" Every student in class reports a response to the prompt. Teachers need to develop a mechanism by which poll results are collected and displayed so that this frequent form of classroom interaction is seamless. Smaller classes might be able to conduct a roll call by voice. For larger classes, students might be given mini-whiteboards or a set of colored index cards to hold up. Free and paid commercial tools are also readily available.

- **Why** Collecting data from the class to use in an activity makes the outcome of the activity more interesting. In other cases, going on record with an estimate makes people want to know if they were right and increases investment in the outcome. If coming up with an estimate is too daunting, ask students for a guess that they are sure is too low or too high. Putting some boundaries on possible outcomes of a problem is an important skill for mathematical modeling (MP4).

Take Turns

- **What** Students work with a partner or small group. They take turns in the work of the activity, whether it be spotting matches, explaining, justifying, agreeing or disagreeing, or asking clarifying questions. If they disagree, they are expected to support their case and listen to their partner's arguments. The first few times students engage in these activities, the teacher should demonstrate, with a partner, how the discussion is expected to go. Once students are familiar with these structures, less set-up will be necessary. While students are working, the teacher can ask students to restate their question more clearly or paraphrase what their partner said.

- **Why** Building in an expectation, through the routine, that students explain the rationale for their choices and listen to another's rationale deepens the understanding that can be achieved through these activities. Specifying that students take turns deciding, explaining, and listening limits the phenomenon where one student takes over and the other does not participate. Taking turns can also give students more opportunities to construct logical arguments and critique others' reasoning (MP3).

Think-Pair-Share

- **What** Students have quiet time to think about a problem and work on it individually, and then time to share their response or their progress with a partner. Once these partner conversations have taken place, some students are selected to share their thoughts with the class.

- **Why** This is a teaching routine useful in many contexts whose purpose is to give all students enough time to think about a prompt and form a response before they are expected to try to verbalize their thinking. First they have an opportunity to share their thinking in a low-stakes way with one partner, so that when they share with the class they can feel calm and confident, as well as say something meaningful that might advance everyone's understanding. Additionally, the teacher has an opportunity to eavesdrop on the partner conversations so that she can purposefully select students to share with the class.

Which One Doesn't Belong?

- **What** Students are presented with four figures, diagrams, graphs, or expressions with the prompt "Which one doesn't belong?" Typically, each of the four options "doesn't belong" for a different reason, and the similarities and differences are mathematically significant. Students are prompted to explain their rationale for deciding that one option doesn't belong and given opportunities to make their rationale more precise.

- **Where** Warm Ups
- **Why** Which One Doesn't Belong? fosters a need to define terms carefully and use words precisely (MP6) in order to compare and contrast a group of geometric figures or other mathematical representations.

Mathematical Modeling Prompts

Mathematics is a tool for understanding the world better and making decisions. School mathematics instruction often neglects giving students opportunities to understand this, and reduces mathematics to disconnected rules for moving symbols around on paper. Mathematical modeling is the process of choosing and using appropriate mathematics and statistics to analyze empirical situations, to understand them better, and to improve decisions (NGA 2010). This mathematics will remain important beyond high school in students' lives and education after high school (NCEE 2013).

The mathematical modeling prompts and this guidance for how to use them represent our effort to make authentic modeling accessible to all teachers and students using this curriculum.

Organizing Principles about Mathematical Modeling

- The purpose of mathematical modeling in school mathematics courses is for students to understand that they can use math to better understand things they are interested in in the world.
- Mathematical modeling is different from solving word problems. It often feels like initially you are not given enough information to answer the question. There should be room to interpret the problem. There ought to be a range of acceptable assumptions and answers. Modeling requires genuine choices to be made by the modeler.
- It is expected that students have support from their teacher and classmates while modeling with mathematics. It is not a solitary activity. Assessment should focus on feedback that helps students improve their modeling skills.

Things the Modeler Does When Modeling with Mathematics (NGA 2010)

1. **Pose a problem** that can be explored with quantitative methods. Identify variables in the situation and select those that represent essential features.
2. **Formulate a model:** create and select geometric, graphical, tabular, algebraic, or statistical representations that describe relationships between variables.
3. **Compute:** Analyze these relationships and perform computations to draw conclusions.
4. **Interpret** the conclusions in terms of the original situation.
5. **Validate** the conclusions by comparing them with the situation. Iterate if necessary to improve the model.
6. **Report** the conclusions and the reasoning behind them.

It's important to recognize that in practice, these actions don't often happen in a nice, neat order.

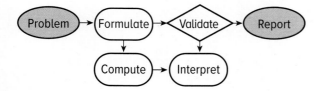

When to Use Mathematical Modeling Prompts

A component of this is mathematical modeling prompts. Prompts include multiple versions of a task (the multiple versions require students engage in more or fewer aspects of mathematical modeling), sample solutions, instructions to teachers for launching the prompt in class and supporting students with that particular prompt, and an analysis of each version showing how much of a "lift" the prompt is along several dimensions of mathematical modeling. A mathematical modeling prompt could be done as a classroom lesson or given as a project. This is a choice made by the teacher.

A mathematical modeling prompt done as a classroom lesson could take one day of instruction or more than one day, depending on how much of the modeling cycle students are expected to engage in, how extensively they are expected to revise their model, and how elaborate the reporting requirements are.

A mathematical modeling prompt done as a project could span several days or weeks. The project is assigned and students work on it in the background while daily math lessons continue to be conducted. (Much like research papers or creative writing assignments in other content areas.) This structure has the advantage of giving students extended time for more complex modeling prompts that would not be feasible to complete in one class period and affords more time for iterations on the model and cycles of feedback.

Modeling prompts don't necessarily need to involve the same math as the current unit of study. As such, the prompts can be given at any time as long as students have the background to construct a reasonable model.

Students might flex their modeling muscles using mathematical concepts that are below grade level. First of all, learning to model mathematically is demanding—learning to do it while also learning new math concepts is likely to be out of reach. Second of all, we know that in future life and work, when students will be called on to engage in mathematical modeling, they will often need to apply math concepts from early grades to ambiguous situations (Forman & Steen, 1995). This elusive category of problems which are high school level yet draw on mathematics first learned in earlier grades may seem contradictory in a curriculum that takes focus and alignment seriously. However, p. 84 of the standards alludes to such problems, and Table 1 in the high school publisher's criteria (p. 8) leaves room for including such problems in high school materials in column 6.

The mathematical modeling prompts are not the only opportunities for students to engage in aspects of mathematical modeling in the curriculum. Mathematical modeling is often new territory for both students and teachers. Oftentimes within the regular classroom lessons, activities include scaled-back modeling scenarios, for which students only need to engage in a part of the modeling cycle. These activities are tagged with the "Aspects of Modeling" instructional routine, and the specific opportunity to engage in an aspect of modeling is explained in the activity narrative.

How to Prepare and Conduct the Modeling Lesson or Project

- Decide which version of the prompt to give.
- Have data ready to share if you plan to give it when students ask.
- Ensure students have access to tools they might be expected to use
- If desired, instruct students to use a template for organizing modeling work.
- Whether doing the prompt as a classroom lesson or giving as a project, plan to do the in-class launch in class.
- Decide to what extent students are expected to iterate and refine their model.
  - If you are conducting a one-day lesson, students may not have much time to refine their model and may not engage as much in that part of the modeling cycle.
  - If you conduct a lesson that takes more than one day, or give the task as a project, it is more reasonable to expect students to iterate and refine their model once or even several times.
- Decide how students will report their results.
  - If conducting a one-day lesson, this may be a rough visual display on a whiteboard.
  - If more time is allotted or the task is assigned as a project, you might instruct students to write a more formal report, slideshow, blog post, poster, or create an a mockup of an artifact like a letter to a specific audience, a smartphone app, a menu, or a set of policies for a government entity to consider.
    - One way to scaffold this work is to ask students to turn in a certain number of presentation slides: one that states the assumptions made, one that describes the model, and one or more slides with their conclusions or recommendations.
- Decide how students will be assessed. Prepare a rubric that will be used and share it with them.

Ideas for Setting Up an Environment Conducive to Modeling

- Provide plenty of blank whiteboard or chalkboard space for groups to work together comfortably. "Vertical non-permanent surfaces" are most conducive to productive collaborative work. "Vertical" means on a vertical wall is better than horizontally on a tabletop, and "non-permanent" means something like a dry erase board is better than something like chart paper (Liljedahl, 2016).
- Ensure that students have easy access to any tools that might be useful for the task. These might include:
  - A supply table containing geometry tools, calculators, scratch paper, graph paper, dry erase markers, post-its
  - Electronic devices to access digital tools (like graphing technology, dynamic geometry software, or statistical technology)
- Think about how you will help students manage the time that is available to work on the task. For example:
  - For lessons, display a countdown timer for intermittent points in the class when you will ask each group to summarize their progress
  - For lessons, decide what time you will ask groups to transition to writing down their findings in a somewhat organized way (perhaps 15 minutes before the end of the class)
  - For projects, set some intermediate milestone deadlines to help students know if they are on track.

Organizing Students into Teams or Groups

- Mathematical modeling is not a solitary activity. It works best when students have support from each other and their teacher.
- Working with a team can make it possible to complete the work in a finite amount of class time. For example, the team may decide it wants to vary one element of the prompt, and compute the output for each variation. What would be many tedious calculations for one person could be only a few calculations for each team member.
- The members of good modeling groups bring a diverse set of skills and points of view. Scramble the members of modeling teams often, so that students have opportunities to play different roles.

Ways to Support Students While They Work on a Modeling Prompt

- Coach them on ways to organize their work better.
- Provide a template to help them organize their thinking. Over time, some groups may transition away from needing to use a template.
- Remind them of analog and digital tools that are available to them.
- When students get stuck or neglect an important aspect of the work, ask them a question to help them engage more fully in part of the modeling cycle. For example:
  - What quantities are important? Which ones change and which ones stay the same? (identify variables)
  - What information do you know? What information would it be nice to know? How could you get that information? What reasonable assumption could you make? (identify variables)
  - What pictures, diagrams, graphs, or equations might help people understand the relationships between the quantities? (formulate)
  - How are you describing the situation mathematically? Where does your solution come from? (compute)
  - Under what conditions does your model work? When might it not work? (interpret)
  - How could you make your model better? How could you make your model more useful under more conditions? (validate)
  - What parts of your solution might be confusing to someone reading it? How could you make it more clear? (report)

How to Interpret the Provided Analysis of a Modeling Prompt

For any mathematical modeling prompt, different versions are provided. We chose to analyze each version among 5 impactful dimensions that vary the demands on the modeler (OECD, 2013).

Each version of a mathematical modeling prompt is accompanied by an analysis chart that looks like this.

Lift Analysis

Attribute	DQ	QI	SD	AD	M	Avg
Lift	0	1	0	0	2	0.6

Each of the attributes of a modeling problem is scored on a scale from 0–2.

- A lower score indicates a prompt with a "lighter lift": students are engaging in less open, less authentic mathematical modeling.
- A higher score indicates a prompt with a "heavier lift": students are engaging in more open, more authentic mathematical modeling.

This matrix shows the attributes that are part of our analysis of each mathematical modeling prompt. We recognize that not all the attributes have the same impact on what teachers and students do. However, for the sake of simplicity they are all weighted the same when they are averaged.

Index	Attribute	Light Lift (0)	Medium Lift (1)	Heavy Lift (2)
DQ	Defining the Question	well-posed question	• elements of ambiguity • prompt might suggest ways assumptions could be made	• freedom to specify and simplify the prompt • modeler must state assumptions
Q1	Quantities of Interest	key variables are declared	key variables are suggested	key variables are not evident
SD	Source of Data	data is provided	modelers are told what measurements to take or data to look up	modelers must decide what measurements to take or data to look up
AD	Amount of Data Given	modeler is given all the information they need and no more	• some extra information is given and modeler must decide what is important, OR • not enough information is given and modeler must ask for it before teacher provides it	• modeler must sift through lots of given information and decide what is important, OR • not enough information is given and modeler must make assumptions, look it up, or take measurements
M	The Model	a model is given in the form of a mathematical representation	• type of model is suggested in words or by a familiar context, OR • modeler chooses appropriate model from a provided list	careful thought about quantities and relationships or additional work (like constructing a scatterplot or drawing geometric examples) is required to identify type of model to use

We recognize that there are other features of a mathematical modeling prompt that could be varied. In the interests of not making things too complex, we only included 5 dimensions in the lift analysis. However, one might choose to modify a prompt on one of these dimensions:

- whether the scenario is posed with words, a highly-structured image or video, or real-world artifacts like articles or authentic diagrams
- presenting example for student to explore before they are expected to engage with the prompt, versus the prompt suggesting that the modeler generate examples or expecting the modeler to generate examples on their own
- whether the prompt makes decisions about units of measure or expects the modeler to reconcile units of measure or employ dimensional thinking
- whether a pre-made digital or analog tool is provided, instructions given for using a particular tool, use of a particular tool is suggested, or modelers simply have access to familiar tools but are not prompted to use them
- whether a mathematical representation is given, suggested, or modelers have the freedom to select and create representations of their own choosing

References

Consortium for Mathematics and Its Applications (2016). *Guidelines for Assessment and Instruction in Mathematical Modeling Education.* Retrieved November 20, 2017 from http://www.comap.com/Free/GAIMME/index.html

Forman, S. L., & Steen, L. A. (1995). *Mathematics for work and life.* In I. M. Carl (Ed.), Seventy-five years of progress: Prospects for school mathematics (pp. 219–241). Reston, VA: National Council of Teachers of Mathematics.

High School Publisher's Criteria for the Common Core State Standards for Mathematics. Retrieved November 20, 2017 from http://www.corestandards.org/wp-content/uploads/Math_Publishers_Criteria_HS_Spring_2013_FINAL1.pdf

Liljedahl, P. (2016). *Building thinking classrooms: Conditions for problem solving.* In P. Felmer, J. Kilpatrick, & E. Pekhonen (eds.) Posing and Solving Mathematical Problems: Advances and New Perspectives. New York, NY: Springer. Retrieved November 20, 2017 from http://peterliljedahl.com/wp-content/uploads/Building-Thinking-Classrooms-Feb-14-20151.pdf

National Governors Association Center for Best Practices (2010). *Common Core State Standards for Mathematics.*

NCEE (2013). What Does *It Really Mean to Be College and Work Ready?* Retrieved November 20, 2017 from http://ncee.org/college-and-work-ready/

OECD (2013). *Strong Performers and Successful Reformers in Education—Lessons from PISA 2012 for the United States.* Retrieved on November 20, 2017 from http://www.oecd.org/pisa/keyfindings/pisa-2012-results-united-states.htm

Modeling Prompts, Algebra 1 (located in the back of the Teacher Edition and as online printables)

Volume 1	Use any time after Lesson…	Volume 2	Use any time after Lesson…
Evaluating a Sample Response	1-1	Critically Examining National Debt	5-17
Display Your Data	1-15	Designing a Fountain	6-17
A New Heating System	2-23	Planning a Concert	7-2
College Characteristics	3-9		
Giving Bonuses	4-8		
Planning a Vacation	4-18		

Modeling Prompts, Geometry (located in the back of the Teacher Edition and as online printables)

Volume 1	Use any time after Lesson…	Volume 2	Use any time after Lesson…
Evaluating a Sample Response	1-1	A New Container	5-14
The Garden Wall	1-1	Viewing Distances	7-3
2000 Calories	1-9	Is This Fair?	8-6
How Much Water?	2-1		
Scaling a Playground	3-1		
On a Roll	3-11		
So Many Flags	4-9		

Modeling Prompts, Algebra 2 (located in the back of the Teacher Edition and as online printables)

Volume 1	Use any time after Lesson…	Volume 2	Use any time after Lesson…
Evaluating a Sample Response	1-1	Exponential Situations	4-16
Viral Marketing	1-11	Swing Time	5-9
Path of the Planets	2-3	Swept Away	6-19
How Big Is That?	3-2	Do an Experiment	7-3

Supporting Diverse Learners

Supporting English Language Learners

Overview

This curriculum builds on foundational principles for supporting language development for all students. This section aims to provide guidance to help teachers recognize and support students' language development in the context of mathematical sense-making. Embedded within the curriculum are instructional supports and practices to help teachers address the specialized academic language demands in math when planning and delivering lessons, including the demands of reading, writing, speaking, listening, conversing, and representing in math (Aguirre & Bunch, 2012).

Therefore, while these instructional supports and practices can and should be used to support all students learning mathematics, they are particularly well-suited to meet the needs of linguistically and culturally diverse students who are learning mathematics while simultaneously acquiring English.

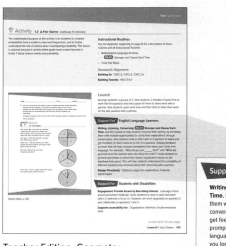

Teacher Edition, Geometry, Lesson 8-1, p. 477

This table reflects the attention and support for language development at each level of the curriculum.

Course	• foundation of curriculum: theory of action and design principles that drive a continuous focus on language development • student glossary of terms
Unit	• unit-specific progression of language development included in each unit overview
Lesson	• language goals embedded in learning goals describe the language demands of the lesson • definitions of new glossary terms
Activity	• additional supports for English language learners based on language demands of the activity • math language routines

Theory of Action

We believe that language development can be built into teachers' instructional practice and students' classroom experience through intentional design of materials, teacher commitments, administrative support, and professional development.

Our theory of action is grounded in the interdependence of language learning and content learning, the importance of scaffolding routines that foster students' independent participation, the value of instructional responsiveness in the teaching process, and the central role of student agency in the learning process.

Mathematical understandings and language competence develop interdependently.

- Deep conceptual learning is gained through language. Ideas take shape through words, texts, illustrations, conversations, debates, examples, etc.
- Teachers, peers, and texts serve as language resources for learning.
- Instructional attention to academic language development, historically limited to vocabulary instruction, has now shifted to also include instruction around the demands of argumentation, explanation, generalization, analyzing the purpose and structure of text, and other mathematical discourse.

Scaffolding provides temporary supports that foster student autonomy.

- Learners with emerging language—at any level—can engage deeply with central mathematical ideas under specific instructional conditions.
- Mathematical language development occurs when students use their developing language to make meaning and engage with challenging problems that are beyond students' mathematical ability to solve independently and therefore require interaction with peers.
- However, these interactions should be structured with temporary supports that students can use to make sense of what is being asked of them, to help organize their own thinking, and to give and receive feedback.

Instruction supports learning when teachers respond to students' verbal and written work.

- Eliciting student thinking through language allows teachers and students to respond formatively to the language students generate. Formative peer and teacher feedback creates opportunities for revision and refinement of both content understandings and language.

Students are agents in their own mathematical and linguistic sense-making.

- Mathematical language proficiency is developed through the process of actively exploring and learning mathematics.
- Language is action: in the very doing of math, students have naturally occurring opportunities to need, learn, and notice mathematical ways of making sense and talking about ideas and the world. These experiences support learners in using as well as expanding their existing language toolkits.

Design

The framework for supporting English language learners (ELLs) in this curriculum includes four design principles for promoting mathematical language use and development in curriculum and instruction. The design principles and related routines work to make language development an integral part of planning and delivering instruction while guiding teachers to amplify the most important language that students are expected to bring to bear on the central mathematical ideas of each unit.

PRINCIPLE 1: Support Sense Making

Scaffold tasks and amplify language so students can make their own meaning.

- Students do not need to understand a language completely before they can engage with academic content in that language. Language learners of all levels can and should engage with grade-level content that is appropriately scaffolded. Students need multiple opportunities to talk about their mathematical thinking, negotiate meaning with others, and collaboratively solve problems with targeted guidance from the teacher.
- Teachers can make language more accessible for students by amplifying rather than simplifying speech or text. Simplifying includes avoiding the use of challenging words or phrases. Amplifying means anticipating where students might need support in understanding concepts or mathematical terms, and providing multiple ways to access them. Providing visuals or manipulatives, demonstrating problem-solving, engaging in think-alouds, and creating analogies, synonyms, or context are all ways to amplify language so that students are supported in taking an active role in their own sense-making of mathematical relationships, processes, concepts, and terms.

PRINCIPLE 2: Optimize Output

Strengthen opportunities and supports for students to describe their mathematical thinking to others, orally, visually, and in writing.

- Linguistic output is the language that students use to communicate their ideas to others (oral, written, visual, etc.), and refers to all forms of student linguistic expressions except those that include significant back-and-forth negotiation of ideas. (That type of conversational language is addressed in the third principle.).
- All students benefit from repeated, strategically optimized, and supported opportunities to articulate mathematical ideas into linguistic expression.
- Opportunities for students to produce output should be strategically optimized for both (a) important concepts of the unit or course, and (b) important disciplinary language functions (for example, making conjectures and claims, justifying claims with evidence, explaining reasoning, critiquing the reasoning of others, making generalizations, and comparing approaches and representations).
  - The focus for optimization must be determined, in part, by how students are currently using language to engage with important disciplinary concepts.
  - When opportunities to produce output are optimized in these ways, students will get spiraled practice in making their thinking about important mathematical concepts stronger with more robust reasoning and examples, and making their thinking clearer with more precise language and visuals.

PRINCIPLE 3: Cultivate Conversation

Strengthen opportunities and supports for constructive mathematical conversations (pairs, groups, and whole class).

- Conversations are back-and-forth interactions with multiple turns that build up ideas about math. Conversations act as scaffolds for students developing mathematical language because they provide opportunities to simultaneously make meaning, communicate that meaning, and refine the way content understandings are communicated.
- When students have a purpose for talking and listening to each other, communication is more authentic. During effective discussions, students pose and answer questions, clarify what is being asked and what is happening in a problem, build common understandings, and share experiences relevant to the topic.
- As mentioned in Principle 2, learners must be supported in their use of language, including when having conversations, making claims, justifying claims with evidence, making conjectures, communicating reasoning, critiquing the reasoning of others, engaging in other mathematical practices, and above all when making mistakes.
- Meaningful conversations depend on the teacher using lessons and activities as opportunities to build a classroom culture that motivates and values efforts to communicate.

PRINCIPLE 4: Maximize Meta-Awareness

Strengthen the meta-connections and distinctions between mathematical ideas, reasoning, and language.

- Language is a tool that not only allows students to communicate their math understanding to others, but also to organize their own experiences, ideas, and learning for themselves.
- Meta-awareness is consciously thinking about one's own thought processes or language use.
- Meta-awareness develops when students and teachers engage in classroom activities or discussions that bring explicit attention to what students need to do to improve communication and reasoning about mathematical concepts.
- When students are using language in ways that are purposeful and meaningful for themselves, in their efforts to understand—and be understood by—each other, they are motivated to attend to ways in which language can be both clarified and clarifying.

These four principles are guides for curriculum development, as well as for planning and execution of instruction, including the structure and organization of interactive opportunities for students. They also serve as guides for and observation, analysis, and reflection on student language and learning. The design principles motivate the use of mathematical language routines, in the *Instructional Routines* section, with examples. The eight routines included in this curriculum are given in the table.

MLR 1:	Stronger and Clearer Each Time
MLR 2:	Collect and Display
MLR 3:	Clarify, Critique, Correct
MLR 4:	Information Gap
MLR 5:	Co-Craft Questions
MLR 6:	Three Reads
MLR 7:	Compare and Connect
MLR 8:	Discussion Supports

Mathematical Language Routines

- When support beyond existing strategies embedded in the curriculum is required, additional supports for English language learners offer instructional strategies for teachers to meet the individual needs of a diverse group of learners.
- Lesson- and activity-level supports for English language learners stem from the design principles and are aligned to the language domains of reading, writing, speaking, listening, conversing, and representing in math (Aguirre & Bunch, 2012).
- These lesson-specific supports provide students with access to the mathematics by supporting them with the language demands of a specific activity without reducing the mathematical demand of the task.
- Using these supports will help maintain student engagement in mathematical discourse and ensure that the struggle remains productive.
- All of the supports are designed to be used as needed, and use should be faded out as students develop understanding and fluency with the English language.
- Teachers should use their professional judgment about which supports to use and when, based on their knowledge of the individual needs of students in their classroom.
- A teacher who notices that students' written responses could get stronger and clearer with more opportunity to revise their writing could use this support to provide students with multiple opportunities to gain additional input, through direct and indirect feedback from their peers.
- Based on their observations of student language, teachers can make adjustments to their teaching and provide additional language support where necessary.
- Teachers can select from the Supports for English language learners provided in the curriculum as appropriate. When selecting from these supports, teachers should take into account the language demands of the specific activity and the language needed to engage the content more broadly, in relation to their students' current ways of using language to communicate ideas as well as their students' English language proficiency.

The *mathematical language routines* were selected because they are effective and practical for simultaneously learning mathematical practices, content, and language.

- A mathematical language routine is a structured but adaptable format for amplifying, assessing, and developing students' language.
- The routines emphasize uses of language that is meaningful and purposeful, rather than about just getting answers.
- These routines can be adapted and incorporated across lessons in each unit to fit the mathematical work wherever there are productive opportunities to support students in using and improving their English and disciplinary language use.
- These routines facilitate attention to student language in ways that support in-the-moment teacher-, peer-, and self-assessment.
- The feedback enabled by these routines will help students revise and refine not only the way they organize and communicate their own ideas, but also ask questions to clarify their understandings of others' ideas.

For a description of each MLR, see the Instructional Routines section.

Supporting Students with Disabilities

All students are individuals who can know, use, and enjoy mathematics. These materials empower students with activities that capitalize on their existing strengths and abilities to ensure that all learners can participate meaningfully in rigorous mathematical content. Lessons support a flexible approach to instruction and provide teachers with options for additional support to address the needs of a diverse group of students.

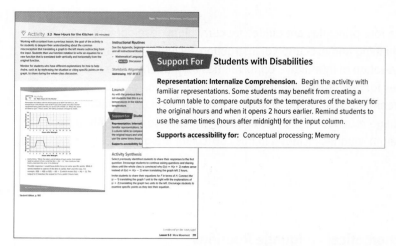

Teacher Edition, Algebra 2, Lesson 5-3, p. 211

Curriculum Features that Support Access

Each lesson is carefully designed to maximize engagement and accessibility for all students. Purposeful design elements that support all learners, but that are especially helpful for students with disabilities include:

Lesson Structures are Consistent

The structure of every lesson is the same: warm up, activities, synthesis, cool down. By keeping the components of each lesson similar from day to day, the flow of work in class becomes predictable for students. This reduces cognitive demand and enables students to focus on the mathematics at hand rather than the mechanics of the lesson.

Concepts Develop from Concrete to Abstract

Mathematical concepts are introduced simply, concretely, and repeatedly, with complexity and abstraction developing over time. Students begin with concrete examples, and transition to diagrams and tables before relying exclusively on symbols to represent the mathematics they encounter.

Individual to Pair or Small Group to Whole Class Progression

Providing students with time to think through a situation or question independently before engaging with others, allows students to carry the weight of learning, with supports arriving just in time from the community of learners. This progression allows students to first activate what they already know, and continue to build from this base with others.

Opportunities to Apply Mathematics to Real-World Contexts

Giving students opportunities to apply the mathematics they learn clarifies and deepens their understanding of core math concepts and skills and provides motivation and support. Mathematical modeling is a powerful activity for all students, but especially students with disabilities. Each unit has a culminating activity designed to explore, integrate, and apply all the big ideas of the unit. Centering instruction on these contextual situations can provide students with disabilities an anchor with which to base their mathematical understandings.

Instructional Strategies that Support Access

The following general instructional strategies can be used to make activities accessible to all students:

Eliminate Barriers
Eliminate any unnecessary barriers that students may encounter that prevent them from engaging with the important mathematical work of a lesson. This requires flexibility and attention to areas such as the physical environment of the classroom, access to tools, organization of lesson activities, and means of communication.

Processing Time
Increased time engaged in thinking and learning leads to mastery of grade-level content for all students, including students with disabilities. Frequent switching between topics creates confusion and does not allow for content to deeply embed in the mind of the learner. Mathematical ideas and representations are carefully introduced in the materials in a gradual, purposeful way to establish a base of conceptual understanding. Some students may need additional time, which should be provided as required.

Assistive Technology
Assistive technology can be a vital tool for students with learning disabilities, visual spatial needs, sensory integration, and students with autism. Assistive technology supports suggested in the materials are designed to either enhance or support learning, or to bypass unnecessary barriers.

Manipulatives
Physical manipulatives help students make connections between concrete ideas and abstract representations. Often, students with disabilities benefit from hands-on activities, which allow them to make sense of the problem at hand and communicate their own mathematical ideas and solutions.

Visual Aids
Visual aids such as images, diagrams, vocabulary anchor charts, color coding, or physical demonstrations, are suggested throughout the materials to support conceptual processing and language development. Many students with disabilities have working memory and processing challenges. Keeping visual aids visible on the board allows students to access them as needed, so that they can solve problems independently. Leaving visual aids on the board especially benefit students who struggle with working or short term memory issues.

Graphic Organizers
Word webs, Venn diagrams, tables, and other metacognitive visual supports provide structures that illustrate relationships between mathematical facts, concepts, words, or ideas. Graphic organizers can be used to support students with organizing thoughts and ideas, planning problem solving approaches, visualizing ideas, sequencing information, or comparing and contrasting ideas.

Brain Breaks
Brain breaks are short, structured, 2–3 minute movement breaks taken in between activities, or to break up a longer activity (approximately every 20–30 minutes during a class period). Brain breaks are a quick, effective way of refocusing and re-energizing the physical and mental state of students during a lesson. Brain breaks have also been shown to positively impact student concentration and stress levels, resulting in more time spent engaged in mathematical problem solving. This universal support is beneficial for all students, but especially those with ADHD.

Supports for Students with Disabilities

The additional supports for students with disabilities are activity-specific and provide teachers with strategies to increase access and eliminate barriers without reducing the mathematical demand of the task. Designed for students with disabilities, they are also appropriate for many students who need additional support to access rigorous, grade-level content. In addition to the guidance provided here, teachers should consider the individual needs of their students and use formative assessment to determine which supports to use and when.

Students' Strengths and Needs

Students' strengths and needs in the following areas of cognitive functioning are integral to learning mathematics (Brodesky et al., 2002) and provide an additional lens to help teachers select appropriate supports for specific types of learner needs.

- **Conceptual Processing** includes perceptual reasoning, problem solving, and metacognition.
- **Language** includes auditory and visual language processing and expression.
- **Visual-Spatial Processing** includes processing visual information and understanding relation in space (e.g., visual mathematical representations and geometric concepts).
- **Organization** includes organizational skills, attention, and focus.
- **Memory** includes working memory and short-term memory.
- **Social-Emotional Functioning** includes interpersonal skills and the cognitive comfort and safety required in order to take risks and make mistakes.
- **Fine Motor Skills** includes tasks that require small muscle movement and coordination such as manipulating objects (graphing, cutting with scissors, writing).

The additional supports for students with disabilities were designed using the Universal Design for Learning Guidelines (http://udlguidelines.cast.org). Each support aligns to one of the three principles of UDL: engagement, representation, and action and expression.

Engagement

Students' attitudes, interests, and values help to determine the ways in which they are most engaged and motivated to learn. Supports that align to this principle offer instructional strategies that provide students with multiple means of engagement and include suggestions that, help provide access by...

- leveraging curiosity and students' existing interests,
- leveraging choice around perceived challenge,
- encouraging and supporting opportunities for peer collaboration,
- providing structures that help students maintain sustained effort and persistence during a task, and
- providing tools and strategies designed to help students self-motivate and become more independent.

Representation

Teachers can reduce barriers and leverage students' individual strengths by inviting students to engage with the same content in different ways. Supports that align to this principle offer instructional strategies that provide students with multiple means of representation and include suggestions that offer alternatives for the ways information is presented or displayed, help develop students' understanding and use of mathematical language and symbols, illustrate connections between and across mathematical representations using color and annotations, identify opportunities to activate or supply background knowledge, and describe organizational methods and approaches designed to help students internalize learning.

Action and Expression

Throughout the curriculum, students are invited to share both their understanding and their reasoning about mathematical ideas with others. Supports that align to this principle offer instructional strategies that provide students with multiple means of action and expression and include suggestions that encourage flexibility and choice with the ways students demonstrate their understanding, list sentence frames that support discourse or accompany writing prompts, indicate appropriate tools, templates, and assistive technologies, support the development of organizational skills in problem-solving, and provide checklists that enable students to monitor their own progress.

For additional information about the Universal Design for Learning Framework, or to learn more about supporting students with disabilities, visit the Center for Applied Special Technology (CAST) at www.cast.org/udl.

References

- Brodesky et al. (2002). Accessibility strategies toolkit for mathematics. Education Development Center. http://www.2.edc.org/accessmath/resources/strategiesToolkit.pdf
- CAST (2018). Universal design for learning guidelines version 2.2. Retrieved from http://udlguidelines.cast.org

Accessibility for Students with Visual Impairments

Features built into the materials that make them more accessible to students with visual impairments include:

1. A color palette using colors that are distinguishable to people with the most common types of color blindness.
2. Tasks and problems designed such that success does not depend on the ability to distinguish between colors.
3. Mathematical diagrams are presented in scalable vector graphs (SVG) format, which can be magnified without loss of resolution, and are possible to render in Braille.
4. Where possible, text associated with images is not part of the image file, but rather, including as an image caption that is accessible to screen readers.
5. Alt text on all images, to make the materials easier to interpret to users accessing the materials with a screen reader.

If students with visual impairments are accessing the materials using a screen reader, it is important to understand:

- All images in the curriculum have alt text: a very short indication of the image's contents, so that the screen reader doesn't skip over as if nothing is there.
- Some images have a longer description to help a student with a visual impairment recreate the image in their mind.

It is important for teachers to understand that students with visual impairments are likely to need help accessing images in lesson activities and assessments, and prepare appropriate accommodations. Be aware that mathematical diagrams are provided as scalable vector graphics (SVG) formats, because this format can be magnified without loss of resolution.

Accessibility experts who reviewed this curriculum recommended that students who would benefit should have access to a Braille version of the curriculum materials, because a verbal description of many of the complex mathematical diagrams would be inadequate for supporting their learning. All diagrams are provided in the SVG file type so that they can be rendered in Braille format.

Assessments

Learning Goals and Targets

Learning Goals
Teacher-facing learning goals appear at the top of lesson plans. They describe, for a teacher audience, the mathematical and pedagogical goals of the lesson.

Student-facing learning goals appear in student materials at the beginning of each lesson and start with the word "Let's." They are intended to invite students into the work of that day without giving away too much and spoiling the problem-based instruction. They are suitable for writing on the board before class begins.

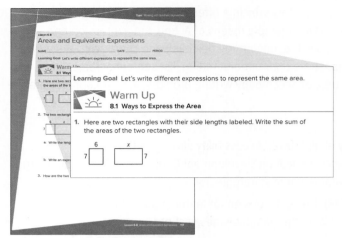

Student Edition, Algebra 1 Support0, Lesson 6-8, p.117

Learning Targets
These appear in student materials at the end of each unit. They describe, for a student audience, the mathematical goals of each lesson.

We do not recommend writing learning targets on the board before class begins, because doing so might spoil the problem-based instruction. (The student-facing learning goals (that start with "Let's") are more appropriate for this purpose.)

Teachers and students might use learning targets in a number of ways. Some examples include:
- targets for standards-based grading
- prompts for a written reflection as part of a lesson synthesis
- a study aid for self-assessment, review, or catching up after an absence from school

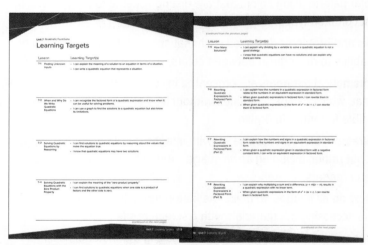

Student Edition, Algebra 1, Unit 7, pp. LT13-14

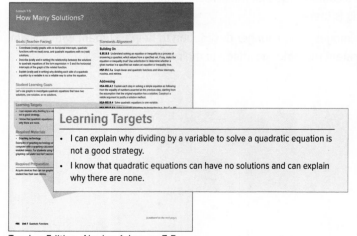

Teacher Edition, Algebra 1, Lesson 7-5, p. 466

44 Assessments

How to Assess Progress

The materials contain many opportunities and tools for both formative and summative assessment. Some things are purely formative, but the tools that can be used for summative assessment can also be used formatively.

Each unit begins with a **diagnostic assessment** (Check Your Readiness) of concepts and skills that are prerequisite to the unit as well as a few items that assess what students already know of the key contexts and concepts that will be addressed by the unit.

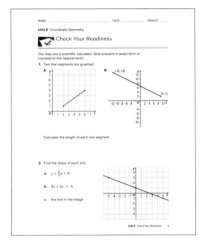

Check Your Readiness, Geometry, Unit 6

Each instructional task is accompanied by commentary about **expected student responses** and **potential misconceptions** so that teachers can adjust their instruction depending on what students are doing in response to the task. Often there are suggested questions to help teachers better understand students' thinking.

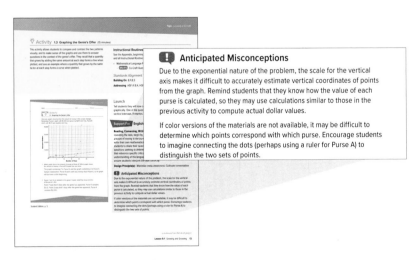

Teacher Edition, Algebra 1, Lesson 5-1, p. 13

Each lesson includes a **cool-down** (analogous to an exit slip or exit ticket) to assess whether students understood the work of that day's lesson. Teachers may use this as a formative assessment to provide feedback or to plan further instruction.

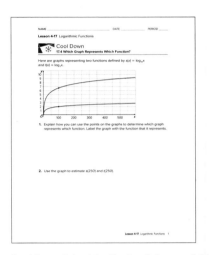

Cool Down Printable, Algebra 2, Lesson 4-17

A set of **practice problems** is provided for each lesson that can be used for homework or in-class practice. The teacher can choose to collect and grade these or simply provide feedback to students.

Each unit includes an **end-of-unit written assessment** that is intended for students to complete individually to assess what they have learned at the conclusion of the unit.

Longer units also include a **mid-unit assessment**. The mid-unit assessment states which lesson in the middle of the unit it is designed to follow.

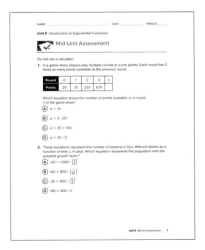

Mid-Unit Assessment Algebra 1, Unit 5

Assessments 45

Diagnostic Assessments
(Check Your Readiness)

At the start of each unit is a *pre-unit diagnostic assessment* called *Check Your Readiness*.

- These assessments vary in length.
- Most of the problems in the pre-unit diagnostic assessment address prerequisite concepts and skills for the unit.
- Teachers can use these problems to identify students with particular below-grade needs, or topics to carefully address during the unit.
- *Check your Readiness* also may include problems that assess what students already know of the upcoming unit's key ideas, which teachers can use to pace or tune instruction; in rare cases, this may signal the opportunity to move more quickly through a topic to optimize instructional time.

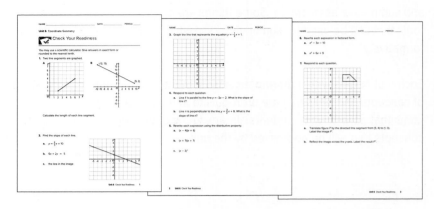

Check Your Readiness, Geometry, Unit 6

What if a large number of students can't do the same pre-unit assessment problem? Teachers are encouraged to address below-grade skills while continuing to work through the on-grade tasks and concepts of each unit, instead of abandoning the current work in favor of material that only addresses below-grade skills.

Look for opportunities within the upcoming unit where the target skill could be addressed in context. For example, an upcoming activity might require solving an equation in one variable. Some strategies might include:

- ask a student who can do the skill to present their method
- add additional questions to the warm-up with the purpose of revisiting the skill
- add to the activity launch a few related equations to solve, before students need to solve an equation while working on the activity
- pause the class while working on the activity to focus on the portion that requires solving an equation

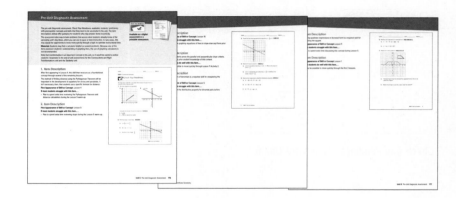

Teacher Edition, Check Your Readiness, Geometry, Unit 6, pp. 175-177

Then, attend carefully to students as they work through the activity. If difficulty persists, add more opportunities to practice the skill, by adapting tasks or practice problems.

What if all students do really well on the pre-unit diagnostic assessment? That means they are ready for the work ahead, and special attention doesn't likely need to be paid to below-grade skill.

Cool Downs

Each lesson includes a cool-down (also known as an exit slip or exit ticket) to be given to students at the end of the lesson. This activity serves as a brief checkpoint to determine whether students understood the main concepts of that lesson. Teachers can use this as a formative assessment to plan further instruction.

Example Cool Down

Here is an example. For a lesson in Algebra 1, Unit 2, the learning goals are

- Create and interpret graphs of inequalities in two variables.
- Write inequalities in two variables to represent situations.

The cool-down reads:

A band is playing at an auditorium with floor seats and balcony seats. The band wants to sell the floor tickets for $15 each and the balcony tickets for $12 each. They want to make at least $3,000 in ticket sales.

1. How much money will they collect for selling floor tickets?
2. How much money will they collect for selling balcony tickets?
3. Write an inequality whose solutions are the number of floor and balcony tickets sold if they make at least $3,000 in ticket sales.
4. Use technology to graph the solutions to your inequality, and sketch the graph.

A number of students were able to write $15x$ to represent the money collected selling floor tickets and $12y$ to represent the money collected selling balcony tickets. However, they wrote the inequality $15x + 12y < 3,000$ and sketched a graph corresponding to this incorrect inequality. You suspect that students interpreted "at least" to mean the same thing as "less than."

Here are some possible strategies.

- In the next four lessons, there are more opportunities to interpret the meaning of "at least." When launching these activities, pause to assist students to interpret this correctly. For example, an activity reads, "He needs at least 9.5 yards of fabric altogether." Ask students, "If he needs at least 9.5 yards, what kinds of numbers are we looking for?" [9.5 yards or anything more than 9.5 yards]. "Since he needs more than 9.5 yards, the amount of fabric needs to be less than 9.5 yards? or greater than 9.5 yards?" [greater than]
- Select the work of one student who answered correctly and one student whose work had the common error. In the next class, display these together for all to see (hide the students' names). Ask each student to decide which interpretation is correct, and defend their choice to their partner. Select students to share their reasoning with the class who have different ways of knowing that "at least $3,000" means "more than $3,000."
- Write feedback for each student along the lines of "What are some different dollar amounts that would satisfy their wish to make at least $3,000?" Allow students to revise and resubmit their work.
- Look for practice problems in upcoming lessons that require students to correctly interpret the term "at least," and be sure to assign those problems.

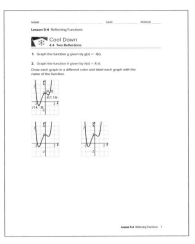

Cool Down Printable, Algebra 2, Lesson 5-4

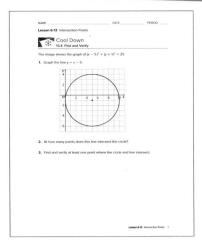

Cool Down Printable, Geometry, Lesson 6-13

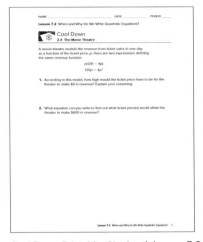

Cool Down Printable, Algebra 1, Lesson 7-2

What if the feedback from a cool-down suggests students haven't understood a key concept?

Choose one or more of these strategies:

- Look at the next few lessons to see if students have more opportunities to engage with the same topic. If so, plan to focus on the topic in the context of the new activities.
- During the next lesson, display the work of a few students on that cool-down. Anonymize their names, but show some correct and incorrect work. Ask the class to observe some things each student did well and could have done better.
- Give each student brief, written feedback on their cool-down that asks a question that nudges them to re-examine their work. Ask students to revise and resubmit.
- Look for practice problems that are similar to, or involve the same topic as the cool-down, then assign those over the next few lessons.

Assessments 47

Summative Assessments

End-of-Unit Assessments

At the end of each unit is the *end-of-unit assessment*. These assessments have a specific length and breadth, with problem types that are intended to gauge students' understanding of the key concepts of the unit while also preparing students for new-generation standardized exams. Problem types include multiple-choice, multiple response, short answer, restricted constructed response, and extended response. Problems vary in difficulty and depth of knowledge.

Teachers may choose to grade these assessments in a standardized fashion, but may also choose to grade more formatively by asking students to show and explain their work on all problems. Teachers may also decide to make changes to the provided assessments to better suit their needs. If making changes, teachers are encouraged to keep the format of problem types provided, which helps students know what to expect and ensures each assessment will take approximately the same amount of time.

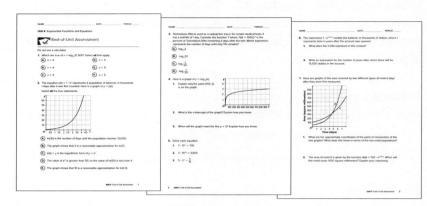

End-of-Unit Assessment, Algebra 2, Unit 4

Mid-Unit Assessments

In longer units, a *mid-unit assessment* is also available. This assessment has the same form and structure as an end-of-unit assessment. In longer units, the end-of-unit assessment will include the breadth of all content for the full unit, with emphasis on the content from the second half of the unit.

All summative assessment problems include a complete solution and standard alignment. Multiple-choice and multiple response problems often include a reason for each potential error a student might make. Restricted constructed response and extended response items include a rubric. Unlike formative assessments, problems on summative assessments generally do not prescribe a method of solution.

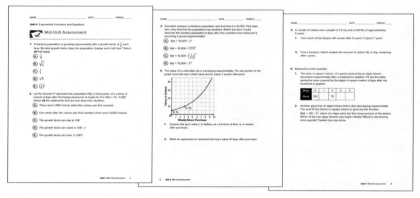

Mid-Unit Assessment, Algebra 2, Unit 4

A note about technology use on assessments: Some assessments require use of technology, some allow it, and some prohibit it. These affordances or restrictions are communicated in each assessment narrative and in instructions to students. Reasons we chose to prohibit use of technology on some assessments include assessing a standard that requires students to sketch a graph by hand, and assessing a standard that requires students to use mathematical properties to rewrite expressions. Conversely, some standards specify that students must use technology for certain things, like generating a best-fit line and correlation coefficient. On assessments where these skills are assessed, technology is required. This approach is in keeping with many state and national standardized assessments that include calculator-allowed and calculator-prohibited portions. Our approach, though, is to allow or prohibit technology on an entire assessment—no single assessment in this curriculum contains both technology-allowed and technology-prohibited portions.

Design Principles for Summative Assessments

Students should get the correct answer on assessment problems for the right reasons, and get incorrect answers for the right reasons. To help with this, our assessment problems are targeted and short, use consistent, positive wording, and have clear, undebatable correct responses.

Multiple-Choice Problems

In multiple choice problems, distractors are common errors and misconceptions directly relating to what is being assessed, since problems are intended to test whether the student has proficiency on a specific skill.

- The distractors serve as a diagnostic, giving teachers the chance to quickly see which of the most common errors are being made.
- There are no "trick" questions, and the phrases "all of the above" and "none of the above" are never used, since they do not give useful information about the methods a student used.

End-of-Unit Assessment, Algebra 1,
Unit 7, Problem 3

Multiple-Response Prompts

Multiple response prompts always include the phrase "select **all**" to clearly indicate their type. Each part of a multiple response problem addresses a different piece of the same overall skill, again serving as a diagnostic for teachers to understand which common errors students are making.

End-of-Unit Assessment, Algebra 1,
Unit 7, Problem 2

Short Answer, Constructed Response, Extended Response

Short answer, restricted constructed response, and extended response problems are careful to avoid the "double whammy" effect, where a part of the problem asks for students to use correct work from a previous part. This choice is made to ensure that students have all possible opportunities to show proficiency on assessments.

End-of-Unit Assessment, Algebra 1,
Unit 7, Problem 7

Extended Response

When possible, extended response problems provide multiple ways for students to demonstrate understanding of the content being assessed, through some combination of arithmetic or algebra, use of representations (tables, graphs, diagrams, expressions, and equations) and explanation.

End-of-Unit Assessment, Algebra 1,
Unit 7, Problem 6

Rubrics for Evaluating Student Answers

Restricted constructed response and extended response items have rubrics that can be used to evaluate the level of student responses.

Restricted Constructed Response

- *Tier 1 response*: Work is complete and correct.
- *Tier 2 response*: Work shows general conceptual understanding and mastery, with some errors.
- *Tier 3 response*: Significant errors in work demonstrate lack of conceptual understanding or mastery. Two or more error types from Tier 2 response can be given as the reason for a Tier 3 response instead of listing combinations.

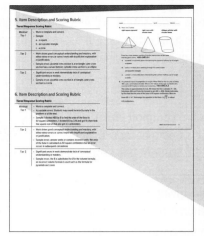

Teacher Edition, Geometry, Unit 5,
End-of-Unit Assessment, p. 170, Problem 5

Extended Response

- *Tier 1 response*: Work is complete and correct, with complete explanation or justification.
- *Tier 2 response*: Work shows good conceptual understanding and mastery, with either minor errors or correct work with insufficient explanation or justification.
- *Tier 3 response*: Work shows a developing but incomplete conceptual understanding, with significant errors.
- *Tier 4 response*: Work includes major errors or omissions that demonstrate a lack of conceptual understanding and mastery.

Typically, sample errors are included. Acceptable errors can be listed at any Tier (as an additional bullet point), notably Tier 1, to specify exclusions.

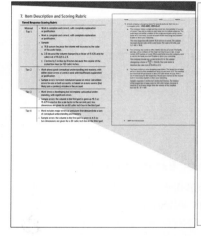

Teacher Edition, Geometry, Unit 5,
End-of-Unit Assessment, p. 171, Problem 7

Curriculum Pacing Guide

Lessons and Assessments Only (*not* including time for modeling prompts or optional lessons)

Week	Algebra 1	Geometry	Algebra 2
1	**Unit 1** One-variable Statistics 13 Days Optional Lessons: 2, 5, 6, 7, 8	**Unit 1** (MA) Constructions and Rigid Transformations 25 Days Optional Lessons: 8, 18, 22	**Unit 1** Sequences and Functions 11 Days Optional Lessons: 4, 6
2			
3			
4	**Unit 2** (MA) Linear Equations, Inequalities, and Systems 29 Days Optional Lessons: none		**Unit 2** (MA) Polynomials and Rational Functions 29 Days Optional Lessons: none
5			
6		**Unit 2** Congruence 16 Days Optional Lesson: 11	
7			
8			
9		**Unit 3** Similarity 15 Days Optional Lessons: 2, 10, 12	**Unit 3** Complex Numbers and Rational Exponents 14 Days Optional Lessons: 1, 2, 9, 14, 16, 19
10	**Unit 3** Two-variable Statistics 11 Days Optional Lesson: 10		
11			
12		**Unit 4** Right Triangle Trigonometry 11 Days Optional Lessons: 2, 3	**Unit 4** (MA) Exponential Functions and Equations 19 Days Optional Lessons: 2, 18
13	**Unit 4** (MA) Functions 21 Days Optional Lessons: none		
14			
15		**Unit 5** Solid Geometry 20 Days Optional Lessons: none	
16			**Unit 5** Transformations of Functions 13 Days Optional Lessons: none
17	**Unit 5** (MA) Introduction to Exponential Functions 23 Days Optional Lesson: 14		
18			
19		**Unit 6** Coordinate Geometry 18 Days Optional Lessons: none	**Unit 6** (MA) Trigonometric Functions 22 Days Optional Lessons: none
20			
21			
22	**Unit 6** (MA) Introduction to Quadratic Functions 19 Days Optional Lesson: 13		
23		**Unit 7** Circles 14 Days Optional Lessons: none	
24			**Unit 7** (MA) Statistical Inferences 18 Days Optional Lesson: 4
25			
26	**Unit 7** (MA) Quadratic Equations 28 Days Optional Lessons: none	**Unit 8** Conditional Probability 11 Days Optional Lessons: 1, 11	
27			
28			
29			
30			

(MA) = Unit has Mid-Unit Assessment

Total number of days for each course = Lessons + Assessments − Optional Lessons

Algebra 1 = 144 days   Geometry = 130 days   Algebra 2 = 126 days

Unit Dependency Chart, High School

In the unit dependency chart shown, an arrow indicates that a particular unit is designed for students who already know the material in a previous unit. Reversing the order would have a negative effect on mathematical or pedagogical coherence.

Here are some examples.

Algebra 1	Geometry
Some are a consequence of choices made by the authors. For example, there is an arrow from A1.5 to A1.6, because when quadratic functions are introduced, they are contrasted with exponential functions, assuming that students are already familiar with exponential functions.	Some dependencies are needed to address the standards. For example, there is an arrow from G.3 to G.4, because students learn that by similarity, side ratios in right triangles are properties of the angles in the triangle, leading to definitions of trigonometric ratios for acute angles.

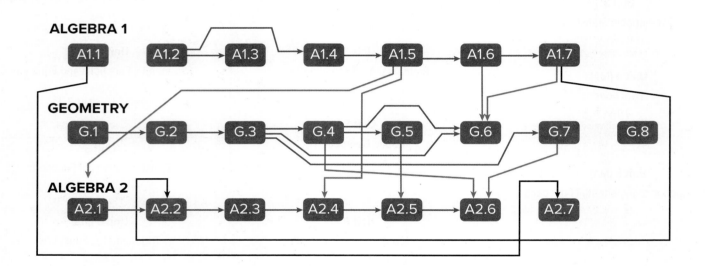

Required Materials, High School

Algebra 1
- ☐ A collection of balls that bounce
- ☐ Blank paper
- ☐ Colored pencils
- ☐ Copies of blackline master
- ☐ Four-function calculators
- ☐ Glue or glue sticks
- ☐ Graphing technology - Examples of graphing technology are: a handheld graphing calculator, a computer with a graphing calculator application installed, and an internet-enabled device with access to a site like desmos.com/calculator or geogebra.org/graphing. For students using the digital materials, a separate graphing calculator tool isn't necessary; interactive applets are embedded throughout, and a graphing calculator tool is accessible on the student digital toolkit page.
- ☐ Graph paper
- ☐ Internet-enabled device
- ☐ Measuring tapes
- ☐ Pre-printed cards, cut from copies of the blackline master
- ☐ Pre-printed slips, cut from copies of the blackline master
- ☐ Rulers
- ☐ Scientific calculators
- ☐ Scissors
- ☐ Spreadsheet technology
- ☐ Statistical technology
- ☐ Sticky notes
- ☐ Tools for creating a visual display - Any way for students to create work that can be easily displayed to the class. Examples: chart paper and markers, whiteboard space and markers, shared online drawing tool, access to a document camera.

Algebra 1 Extra Support Materials
- ☐ Coins - any fair two-sided coin
- ☐ Index cards
- ☐ Masking tape
- ☐ Meter sticks
- ☐ Number cubes – cubes with sides numbered from 1 to 6
- ☐ Pennies
- ☐ Slips of paper
- ☐ Tape measures

Geometry

- ☐ 1-inch strips cut from card stock with evenly-spaced holes - Copy the two sizes of strips on different colors of card stock, so that students can see at a glance whether the two strips being used are congruent or not. After the strips are cut, punch holes as indicated in each segment. A standard hole punch makes holes that are a little larger than needed for the metal paper fasteners, causing the cardboard strips to wiggle around. If possible, find a way to punch holes that are slightly smaller than the size of a standard hole punch. A commercial version of these materials is available, sometimes called geo strips.
- ☐ Blank paper
- ☐ Clay
- ☐ Coins - any fair two-sided coin
- ☐ Colored pencils
- ☐ Copies of blackline master
- ☐ Cylindrical food items - Approximately cylindrical food items that can be easily sliced. Examples include carrots or cheese sticks.
- ☐ Data collected from the previous activity
- ☐ Dental floss
- ☐ Dried linguine pasta - We specified linguine since it is flatter and less likely to roll around than spaghetti.
- ☐ Dynamic geometry software
- ☐ Four-function calculators
- ☐ $\frac{1}{2}$-inch cubes
- ☐ Geometry toolkits (HS) - Index cards to use as straightedges, compasses, tracing paper, blank paper, colored pencils, and scissors. Note: "Tracing paper" is easiest to use when it's a smaller size. Commercially-available "patty paper" is 5 inches by 5 inches and ideal for this. If using larger sheets of tracing paper, consider cutting them down for student use.
- ☐ Graphing technology - Examples of graphing technology are: a handheld graphing calculator, a computer with a graphing calculator application installed, and an internet-enabled device with access to a site like desmos.com/calculator or geogebra.org/graphing. For students using the digital materials, a separate graphing calculator tool isn't necessary; interactive applets are embedded throughout, and a graphing calculator tool is accessible on the student digital toolkit page.
- ☐ Graph paper
- ☐ Highlighters
- ☐ Index cards
- ☐ Internet-enabled device
- ☐ Isometric dot paper
- ☐ Masking tape
- ☐ Measuring tapes
- ☐ Measuring tools
- ☐ Metal paper fasteners - brass brads
- ☐ Mirrors - Flat mirrors. A mirror should be small enough to be portable, but large enough that it can be placed on the ground a few feet away from an observer, and the observer can site the top of a tall object in the mirror.
- ☐ Number cubes - cubes with sides numbered from 1 to 6
- ☐ Origami paper
- ☐ Paper bags
- ☐ Pre-cut figures - Figures cut from cardboard or card stock with at least one straight side that can be taped to a pencil. Shapes might include various polygons, half-discs, and composites of these.
- ☐ Pre-printed cards, cut from copies of the blackline master
- ☐ Pre-printed slips, cut from copies of the blackline master
- ☐ Protractors - Clear protractors with no holes and with radial lines printed on them are recommended.
- ☐ Rulers
- ☐ Rulers marked with centimeters
- ☐ Scientific calculators
- ☐ Scissors
- ☐ Spreadsheet technology
- ☐ Sticky notes
- ☐ Straws
- ☐ String
- ☐ Tape
- ☐ Tools for creating a visual display - Any way for students to create work that can be easily displayed to the class. Examples: chart paper and markers, whiteboard space and markers, shared online drawing tool, access to a document camera.
- ☐ Unlined index cards

Algebra 2
- ☐ Blank paper
- ☐ Circular objects of different sizes - A collection of objects that are a circle or have a circular face. Examples include plates, cans, cookie tins, and hockey pucks.
- ☐ Coins - any fair two-sided coin
- ☐ Colored pencils
- ☐ Copies of blackline master
- ☐ Graphing technology - Examples of graphing technology are: a handheld graphing calculator, a computer with a graphing calculator application installed, and an internet-enabled device with access to a site like desmos.com/calculator or geogebra.org/graphing. For students using the digital materials, a separate graphing calculator tool isn't necessary; interactive applets are embedded throughout, and a graphing calculator tool is accessible on the student digital toolkit page.
- ☐ Graph paper
- ☐ Index cards
- ☐ Internet-enabled device
- ☐ Number cubes cubes with sides numbered from 1 to 6
- ☐ Objects for Tower of Hanoi - If not using the digital applet, each student could use a quarter, nickel, penny, and dime, and a piece of paper with 3 circles drawn on it.
- ☐ Paper bags
- ☐ Pre-printed slips, cut from copies of the blackline master
- ☐ Random number generator
- ☐ Ribbon or string
- ☐ Rulers
- ☐ Rulers marked with centimeters
- ☐ Scientific calculators
- ☐ Scissors
- ☐ Slips of paper
- ☐ Slips with student names
- ☐ Spreadsheet technology
- ☐ Statistical technology
- ☐ Stopwatches
- ☐ Straightedges - A rigid edge that can be used for drawing line segments. Sometimes a ruler is okay to use as a straightedge, but sometimes it is preferable to use an unruled straightedge, like a blank index card.
- ☐ Tape
- ☐ Tools for creating a visual display - Any way for students to create work that can be easily displayed to the class. Examples: chart paper and markers, whiteboard space and markers, shared online drawing tool, access to a document camera.
- ☐ Tracing paper - Bundles of "patty paper" are available commercially for a very low cost. These are small sheets (about 5" by 5") of transparent paper.

Table of Contents, Algebra 1

Refer to the Tables of Contents in the Teacher Edition for more detailed information by unit.

Unit 1 One-variable Statistics

Getting to Know You
- 1-1 Getting to Know You
- 1-2 Data Representations
- 1-3 A Gallery of Data

Distribution Shapes
- 1-4 The Shape of Distributions
- 1-5 Calculating Measures of Center and Variability

How to Use Spreadsheets
- 1-6 Mystery Computations
- 1-7 Spreadsheet Computations
- 1-8 Spreadsheet Shortcuts

Manipulating Data
- 1-9 Technological Graphing
- 1-10 The Effect of Extremes
- 1-11 Comparing and Contrasting Data Distributions
- 1-12 Standard Deviation
- 1-13 More Standard Deviation
- 1-14 Outliers
- 1-15 Comparing Data Sets

Analyzing Data
- 1-16 Analyzing Data

Unit 2 Linear Equations, Inequalities, and Systems

Writing and Modeling with Equations
- 2-1 Planning a Pizza Party
- 2-2 Writing Equations to Model Relationships (Part 1)
- 2-3 Writing Equations to Model Relationships (Part 2)
- 2-4 Equations and Their Solutions
- 2-5 Equations and Their Graphs

Manipulating Equations and Understanding Their Structure
- 2-6 Equivalent Equations
- 2-7 Explaining Steps for Rewriting Equations
- 2-8 Which Variable to Solve for? (Part 1)
- 2-9 Which Variable to Solve for? (Part 2)
- 2-10 Connecting Equations to Graphs (Part 1)
- 2-11 Connecting Equations to Graphs (Part 2)

Systems of Linear Equations in Two Variables
- 2-12 Writing and Graphing Systems of Linear Equations
- 2-13 Solving Systems by Substitution
- 2-14 Solving Systems by Elimination (Part 1)
- 2-15 Solving Systems by Elimination (Part 2)
- 2-16 Solving Systems by Elimination (Part 3)
- 2-17 Systems of Linear Equations and Their Solutions

Linear Inequalities in One Variable
- 2-18 Representing Situations with Inequalities
- 2-19 Solutions to Inequalities in One Variable
- 2-20 Writing and Solving Inequalities in One Variable

Linear Inequalities in Two Variables
- 2-21 Graphing Linear Inequalities in Two Variables (Part 1)
- 2-22 Graphing Linear Inequalities in Two Variables (Part 2)
- 2-23 Solving Problems with Inequalities in Two Variables

Systems of Linear Inequalities in Two Variables
- 2-24 Solutions to Systems of Linear Inequalities in Two Variables
- 2-25 Solving Problems with Systems of Linear Inequalities in Two Variables
- 2-26 Modeling with Systems of Inequalities in Two Variables

Unit 3 Two-variable Statistics

Two-way Tables

3-1 Two-way Tables

3-2 Relative Frequency Tables

3-3 Associations in Categorical Data

Scatterplots

3-4 Linear Models

3-5 Fitting Lines

3-6 Residuals

Correlation Coefficients

3-7 The Correlation Coefficient

3-8 Using the Correlation Coefficient

3-9 Causal Relationships

Estimating Lengths

3-10 Fossils and Flags

Unit 4 Functions

Functions and Their Representations

4-1 Describing and Graphing Situations

4-2 Function Notation

4-3 Interpreting & Using Function Notation

4-4 Using Function Notation to Describe Rules (Part 1)

4-5 Using Function Notation to Describe Rules (Part 2)

Analyzing and Creating Graphs of Functions

4-6 Features of Graphs

4-7 Using Graphs to Find Average Rate of Change

4-8 Interpreting and Creating Graphs

4-9 Comparing Graphs

A Closer Look at Inputs and Outputs

4-10 Domain and Range (Part 1)

4-11 Domain and Range (Part 2)

4-12 Piecewise Functions

4-13 Absolute Value Functions (Part 1)

4-14 Absolute Value Functions (Part 2)

Inverse Functions

4-15 Inverse Functions

4-16 Finding and Interpreting Inverse Functions

4-17 Writing Inverse Functions to Solve Problems

Putting It All Together

4-18 Using Functions to Model Battery Power

Unit 5 Introduction to Exponential Functions

Looking at Growth
- 5-1 Growing and Growing
- 5-2 Patterns of Growth

A New Kind of Relationship
- 5-3 Representing Exponential Growth
- 5-4 Understanding Decay
- 5-5 Representing Exponential Decay
- 5-6 Analyzing Graphs
- 5-7 Using Negative Exponents

Exponential Functions
- 5-8 Exponential Situations as Functions
- 5-9 Interpreting Exponential Functions
- 5-10 Looking at Rates of Change
- 5-11 Modeling Exponential Behavior
- 5-12 Reasoning about Exponential Graphs (Part 1)
- 5-13 Reasoning about Exponential Graphs (Part 2)

Percent Growth and Decay
- 5-14 Recalling Percent Change
- 5-15 Functions Involving Percent Change
- 5-16 Compounding Interest
- 5-17 Different Compounding Intervals
- 5-18 Expressed in Different Ways

Comparing Linear and Exponential Functions
- 5-19 Which One Changes Faster?
- 5-20 Changes over Equal Intervals

Putting It All Together
- 5-21 Predicting Populations

Unit 6 Introduction to Quadratic Functions

A Different Kind of Change
- 6-1 A Different Kind of Change
- 6-2 How Does it Change?

Quadratic Functions
- 6-3 Building Quadratic Functions from Geometric Patterns
- 6-4 Comparing Quadratic and Exponential Functions
- 6-5 Building Quadratic Functions to Describe Situations (Part 1)
- 6-6 Building Quadratic Functions to Describe Situations (Part 2)
- 6-7 **Building Quadratic Functions to Describe Situations (Part 3)**

Working with Quadratic Expressions
- 6-8 Equivalent Quadratic Expressions
- 6-9 Standard Form and Factored Form
- 6-10 Graphs of Functions in Standard and Factored Forms

Features of Graphs of Quadratic Functions
- 6-11 Graphing from the Factored Form
- 6-12 Graphing the Standard Form (Part 1)
- 6-13 Graphing the Standard Form (Part 2)
- 6-14 Graphs That Represent Situations
- 6-15 Vertex Form
- 6-16 Graphing from the Vertex Form
- 6-17 Changing the Vertex

Unit 7 Quadratic Functions

Finding Unknown Inputs

7-1 Finding Unknown Inputs

7-2 When and Why Do We Write Quadratic Equations?

Solving Quadratic Equations

7-3 Solving Quadratic Equations by Reasoning

7-4 Solving Quadratic Equations with the Zero Product Property

7-5 How Many Solutions?

7-6 Rewriting Quadratic Expressions in Factored Form (Part 1)

7-7 Rewriting Quadratic Expressions in Factored Form (Part 2)

7-8 Rewriting Quadratic Expressions in Factored Form (Part 3)

7-9 Solving Quadratic Equations by Using Factored Form

7-10 Rewriting Quadratic Expressions in Factored Form (Part 4)

Completing the Square

7-11 What are Perfect Squares?

7-12 Completing the Square (Part 1)

7-13 Completing the Square (Part 2)

7-14 Completing the Square (Part 3)

7-15 Quadratic Equations with Irrational Solutions

The Quadratic Formula

7-16 The Quadratic Formula

7-17 Applying the Quadratic Formula (Part 1)

7-18 Applying the Quadratic Formula (Part 2)

7-19 Deriving the Quadratic Formula

7-20 Rational and Irrational Solutions

7-21 Sums and Products of Rational and Irrational Numbers

Vertex Form Revisited

7-22 Rewriting Quadratic Expressions in Vertex Form

7-23 Using Quadratic Expressions in Vertex Form to Solve Problems

Putting It All Together

7-24 Using Quadratic Equations to Model Situations and Solve Problems

Table of Contents, Algebra 1 Extra Support Materials

Refer to the Tables of Contents in the Teacher Edition for more detailed information by unit.

Unit 1 One-variable Statistics

Getting to Know You
- 1-1 Human Box Plot
- 1-2 Human Dot Plot
- 1-3 Which One?

Distribution Shapes
- 1-4 The Shape of Data Distributions
- 1-5 Watch Your Calculations

Manipulating Data
- 1-9 Using Technology for Statistics
- 1-10 Measures of Center
- 1-11 Decisions, Decisions
- 1-12 Variability
- 1-13 Standard Deviation in Real-World Contexts
- 1-14 Outliers & Means
- 1-15 Where Are We Eating?

Analyzing Data
- 1-16 Compare & Contrast

Unit 2 Linear Equations, Inequalities and Systems

Writing and Modeling with Equations
- 2-1 Expressing Mathematics
- 2-2 Words and Symbols
- 2-3 Setting the Table
- 2-4 Solutions in Context
- 2-5 Graphs, Tables, and Equations

Manipulating Equations and Understanding Their Structure
- 2-6 Equality Diagrams
- 2-7 Why Is That Okay?
- 2-8 Reasoning About Equations
- 2-9 Same Situation, Different Symbols
- 2-10 Equations and Relationships
- 2-11 Slopes and Intercepts

Systems of Linear Equations in Two Variables
- 2-12 Connecting Situations and Graphs
- 2-13 Making New, True Equations
- 2-14 Making More New, True Equations
- 2-15 Off the Line
- 2-16 Elimination
- 2-17 Number of Solutions in One-Variable Equations

Linear Inequalities in One Variable
- 2-18 Inequalities in Context
- 2-19 Queuing on the Number Line
- 2-20 Interpreting Inequalities

Linear Inequalities in Two Variables
- 2-21 From One- to Two-Variable Inequalities
- 2-22 Situations with Constraints
- 2-23 Modeling Constraints

Systems of Linear Inequalities in Two Variables
- 2-24 Reasoning with Graphs of Inequalities
- 2-25 Representing Systems of Inequalities
- 2-26 Testing Points to Solve Inequalities

Unit 3 Two-variable Statistics

Two-way Tables
- 3-1 Human Frequency Table
- 3-2 Talking Percents
- 3-3 Associations and Relative Frequency Tables

Scatterplots
- 3-4 Interpret This, Interpret That
- 3-5 Goodness of Fit
- 3-6 Actual Data vs. Predicted Data

Correlation Coefficients
- 3-7 Confident Models
- 3-8 Correlations
- 3-9 What's the Correlation?

Estimating Lengths
- 3-10 Putting It All Together

Unit 4 Functions

Functions and Their Representations
- 4-1 Describing Graphs
- 4-2 Understanding Points in Situations
- 4-3 Using Function Notation
- 4-4 Interpreting Functions
- 4-5 Function Representations

Analyzing and Creating Graphs of Functions
- 4-6 Finding Interesting Points on a Graph
- 4-7 Slopes of Segments
- 4-8 Interpreting and Drawing Graphs for Situations
- 4-9 Increasing and Decreasing Functions

A Closer Look at Inputs and Outputs
- 4-10 Interpreting Inputs and Outputs
- 4-11 Examining Domains and Ranges
- 4-12 Functions with Multiple Parts
- 4-13 Number Line Distances
- 4-14 Absolute Value Meaning

Inverse Functions
- 4-15 Finding Input Values and Function Values
- 4-16 Rewriting Equations for Perspectives
- 4-17 Interpreting Function Parts in Situations

Putting It All Together
- 4-18 Modeling Price Information

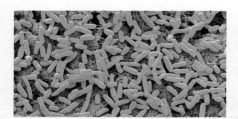

Unit 5 Introduction to Exponential Functions

Looking at Growth
- 5-1 Reviewing Exponents
- 5-2 Growth Patterns

A New Kind of Relationship
- 5-3 Properties of Exponents
- 5-4 Working with Fractions
- 5-5 Connections between Representations
- 5-6 Find That Factor
- 5-7 Negative Exponents

Exponential Functions
- 5-8 Representing Functions
- 5-9 Interpreting Functions
- 5-10 Rate of Change
- 5-11 Skills for Modeling with Mathematics
- 5-12 Connections between Graphs and Equations
- 5-13 Representations of Exponential Functions

Percent Growth and Decay
- 5-14 Percent Increase and Decrease
- 5-15 Changing the Score
- 5-16 Over and Over
- 5-17 Annually, Quarterly, or Monthly?
- 5-18 Bases and Exponents

Comparing Linear and Exponential Functions
- 5-19 Adjusting Windows
- 5-20 Evaluating Functions over Equal Intervals

Putting It All Together
- 5-21 Skills for Mathematical Modeling

Unit 6 Introduction to Quadratic Functions

A Different Kind of Change
- 6-1 Accessing Areas and Pondering Perimeters
- 6-2 Describing Patterns

Quadratic Functions
- 6-3 Lots of Rectangles
- 6-4 Evaluating Quadratic and Exponential Functions
- 6-5 Distance To and Distance From
- 6-6 Graphs of Situations that Change
- 6-7 Accurate Representations

Working with Quadratic Expressions
- 6-8 Areas and Equivalent Expressions
- 6-9 Working with Signed Numbers
- 6-10 Relating Linear Equations and their Graphs

Features of Graphs of Quadratic Functions
- 6-11 Zeros of Functions and Intercepts of Graphs
- 6-12 Changing the Equation
- 6-14 Interpreting Representations
- 6-15 Preparing for Vertex Form
- 6-16 Graphing from the Vertex Form
- 6-17 Parameters and Graphs

Unit 7 Quadratic Functions

Finding Unknown Inputs
7-1 Areas Around Areas
7-2 Equations and Graphs

Solving Quadratic Equations
7-3 Squares and Equations
7-4 Equations and Their Zeros
7-5 Steps in Solving Equations
7-6 Sums and Products
7-7 Integers of Quadratics
7-8 Multiplying Expressions
7-9 Equivalent Equations and Functions
7-10 Quadratic Zeros

Completing the Square
7-11 Finding Perfect Squares
7-12 Forms of Quadratic Equations
7-13 Constants in Quadratic Equations
7-14 Rewriting Quadratic Expressions
7-15 Irrational Numbers

The Quadratic Formula
7-16 Working with Quadratics
7-17 Quadratic Meanings
7-18 Solving Quadratics
7-19 Quadratic Steps
7-20 Quadratics and Irrationals
7-21 Odd and Even Numbers

Vertex Form Revisited
7-22 Features of Parabolas
7-23 Comparing Functions

Putting It All Together
7-24 Quadratic Situations

Table of Contents, Geometry

Refer to the Tables of Contents in the Teacher Edition for more detailed information by unit.

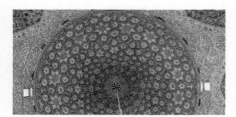

Unit 1 Constructions and Rigid Transformations

Constructions
- 1-1 Build It
- 1-2 Constructing Patterns
- 1-3 Construction Techniques 1: Perpendicular Bisectors
- 1-4 Construction Techniques 2: Equilateral Triangles
- 1-5 Construction Techniques 3: Perpendicular Lines and Angle Bisectors
- 1-6 Construction Techniques 4: Parallel and Perpendicular Lines
- 1-7 Construction Techniques 5: Squares
- 1-8 Using Technology for Constructions
- 1-9 Speedy Delivery

Rigid Transformations
- 1-10 Rigid Transformations
- 1-11 Defining Reflections
- 1-12 Defining Translations
- 1-13 Incorporating Rotations
- 1-14 Defining Rotations
- 1-15 Symmetry
- 1-16 More Symmetry
- 1-17 Working with Rigid Transformations
- 1-18 Practicing Point by Point Transformations

Evidence and Proof
- 1-19 Evidence, Angles, and Proof
- 1-20 Transformations, Transversals, and Proof
- 1-21 One Hundred and Eighty

Designs
- 1-22 Now What Can You Build?

Unit 2 Congruence

Congruent Triangles
- 2-1 Congruent Parts, Part 1
- 2-2 Congruent Parts, Part 2
- 2-3 Congruent Triangles, Part 1
- 2-4 Congruent Triangles, Part 2
- 2-5 Points, Segments, and Zigzags
- 2-6 Side-Angle-Side Triangle Congruence
- 2-7 Angle-Side-Angle Triangle Congruence
- 2-8 The Perpendicular Bisector Theorem
- 2-9 Side-Side-Side Triangle Congruence
- 2-10 Practicing Proofs
- 2-11 Side-Side-Angle (Sometimes) Congruence

Proofs about Quadrilaterals
- 2-12 Proofs about Quadrilaterals
- 2-13 Proofs about Parallelograms
- 2-14 Bisect It

Putting It All Together
- 2-15 Congruence for Quadrilaterals

Unit 3 Similarity

Properties of Dilations

3-1 Scale Drawings

3-2 Scale of the Solar System

3-3 Measuring Dilations

3-4 Dilating Lines and Angles

3-5 Splitting Triangle Sides with Dilation, Part 1

Similarity Transformations and Proportional Reasoning

3-6 Connecting Similarity and Transformations

3-7 Reasoning about Similarity with Transformations

3-8 *Are They All Similar?*

3-9 Conditions for Triangle Similarity

3-10 Other Conditions for Triangle Similarity

3-11 Splitting Triangle Sides with Dilation, Part 2

3-12 Practice With Proportional Relationships

Similarity in Right Triangles

3-13 Using the Pythagorean Theorem and Similarity

3-14 Proving the Pythagorean Theorem

3-15 Finding All the Unknown Values in Triangles

Putting It All Together

3-16 Bank Shot

Unit 4 Right Triangle Trigonometry

Angles and Steepness

4-1 Angles and Steepness

4-2 Half a Square

4-3 Half an Equilateral Triangle

4-4 Ratios in Right Triangles

4-5 Working with Ratios in Right Triangles

Defining Trigonometric Ratios

4-6 Working with Trigonometric Ratios

4-7 Applying Ratios in Right Triangles

4-8 Sine and Cosine in the Same Right Triangle

4-9 Using Trigonometric Ratios to Find Angles

4-10 Solving Problems with Trigonometry

4-11 Approximating Pi

Unit 5 Solid Geometry

Cross Sections, Scaling, and Area

- 5-1 Solids of Rotation
- 5-2 Slicing Solids
- 5-3 Creating Cross Sections by Dilating
- 5-4 Scaling and Area
- 5-5 Scaling and Unscaling

Scaling Solids

- 5-6 Scaling Solids
- 5-7 The Root of the Problem
- 5-8 Speaking of Scaling

Prism and Cylinder Volumes

- 5-9 Cylinder Volumes
- 5-10 Cross Sections and Volume
- 5-11 Prisms Practice

Understanding Pyramid Volumes

- 5-12 Prisms and Pyramids
- 5-13 Building a Volume Formula for a Pyramid
- 5-14 Working with Pyramids
- 5-15 Putting All the Solids Together

Putting It All Together

- 5-16 Surface Area and Volume
- 5-17 Volume and Density
- 5-18 Volume and Graphing

Unit 6 Coordinate Geometry

Transformations in the Plane

- 6-1 Rigid Transformations in the Plane
- 6-2 Transformations as Functions
- 6-3 Types of Transformations

Distances, Circles, and Parabolas

- 6-4 Distances and Circles
- 6-5 Squares and Circles
- 6-6 Completing the Square
- 6-7 Distances and Parabolas
- 6-8 Equations and Graphs

Proving Geometric Theorems Algebraically

- 6-9 Equations of Lines
- 6-10 Parallel Lines in the Plane
- 6-11 Perpendicular Lines in the Plane
- 6-12 It's All on the Line
- 6-13 Intersection Points
- 6-14 Coordinate Proof
- 6-15 Weighted Averages
- 6-16 Weighted Averages in a Triangle

Putting It All Together

- 6-17 Lines in Triangles

Unit 7 Circles

Lines, Angles, and Circles
7-1 Lines, Angles, and Curves
7-2 Inscribed Angles
7-3 Tangent Lines

Polygons and Circles
7-4 Quadrilaterals in Circles
7-5 Triangles in Circles
7-6 A Special Point
7-7 Circles in Triangles

Measuring Circles
7-8 Arcs and Sectors
7-9 Part to Whole
7-10 Angles, Arcs, and Radii
7-11 A New Way to Measure Angles
7-12 Radian Sense
7-13 Using Radians

Putting It All Together
7-14 Putting It All Together

Unit 8 Conditional Probability

Up to Chance
8-1 Up to Chance
8-2 Playing with Probability
8-3 Sample Spaces
8-4 Tables of Relative Frequencies

Combining Events
8-5 Combining Events
8-6 The Addition Rule

Related Events
8-7 Related Events
8-8 Conditional Probability
8-9 Using Tables for Conditional Probability
8-10 Using Probability to Determine Whether Events Are Independent

Conditional Probability
8-11 Probabilities in Games

Table of Contents, Algebra 2

Refer to the Tables of Contents in the Teacher Edition for more detailed information by unit.

Unit 1 Sequences and Functions

A Towering Sequence
- 1-1 A Towering Sequence

Sequences
- 1-2 Introducing Geometric Sequences
- 1-3 Different Types of Sequences
- 1-4 Using Technology to Work with Sequences
- 1-5 Sequences are Functions
- 1-6 Representing Sequences
- 1-7 Representing More Sequences

What's the Equation?
- 1-8 The nth Term
- 1-9 What's the Equation?
- 1-10 Situations and Sequence Types
- 1-11 Adding Up

Unit 2 Polynomials and Rational Functions

What Is a Polynomial?
- 2-1 Let's Make a Box
- 2-2 Funding the Future
- 2-3 Introducing Polynomials
- 2-4 Combining Polynomials

Working with Polynomials
- 2-5 Connecting Factors and Zeros
- 2-6 Different Forms
- 2-7 Using Factors and Zeros
- 2-8 End Behavior (Part 1)
- 2-9 End Behavior (Part 2)
- 2-10 Multiplicity
- 2-11 Finding Intersections
- 2-12 Polynomial Division (Part 1)
- 2-13 Polynomial Division (Part 2)
- 2-14 What Do You Know About Polynomials?
- 2-15 The Remainder Theorem

Rational Functions
- 2-16 Minimizing Surface Area
- 2-17 Graphs of Rational Functions (Part 1)
- 2-18 Graphs of Rational Functions (Part 2)
- 2-19 End Behavior of Rational Functions

Rational Equations
- 2-20 Rational Equations (Part 1)
- 2-21 Rational Equations (Part 2)
- 2-22 Solving Rational Equations

Polynomial Identities
- 2-23 Polynomial Identities (Part 1)
- 2-24 Polynomial Identities (Part 2)
- 2-25 Summing Up
- 2-26 Using the Sum

Unit 3 Complex Numbers and Rational Exponents

Exponent Properties
- 3-1 Properties of Exponents
- 3-2 Square Roots and Cube Roots
- 3-3 Exponents That Are Unit Fractions
- 3-4 Positive Rational Exponents
- 3-5 Negative Rational Exponents

Solving Equations with Square and Cube Roots
- 3-6 Squares and Square Roots
- 3-7 Inequivalent Equations
- 3-8 Cubes and Cube Roots
- 3-9 Solving Radical Equations

A New Kind of Number
- 3-10 A New Kind of Number
- 3-11 Introducing the Number i
- 3-12 Arithmetic with Complex Numbers
- 3-13 Multiplying Complex Numbers
- 3-14 More Arithmetic with Complex Numbers
- 3-15 Working Backwards

Solving Quadratics with Complex Numbers
- 3-16 Solving Quadratics
- 3-17 Completing the Square and Complex Solutions
- 3-18 The Quadratic Formula and Complex Solutions
- 3-19 Real and Non-Real Solutions

Unit 4 Exponential Functions and Equations

Growing and Shrinking
- 4-1 Growing and Shrinking
- 4-2 Representations of Growth and Decay

Understanding Non-Integer Inputs
- 4-3 Understanding Rational Inputs
- 4-4 Representing Functions at Rational Inputs
- 4-5 Changes Over Rational Intervals
- 4-6 Writing Equations for Exponential Functions
- 4-7 Interpreting and Using Exponential Functions

Missing Exponents
- 4-8 Unknown Exponents
- 4-9 What is a Logarithm?
- 4-10 Interpreting and Writing Logarithmic Equations
- 4-11 Evaluating Logarithmic Expressions

The Constant e
- 4-12 The Number e
- 4-13 Exponential Functions with Base e
- 4-14 Solving Exponential Equations

Logarithmic Functions and Graphs
- 4-15 Using Graphs and Logarithms to Solve Problems (Part 1)
- 4-16 Using Graphs and Logarithms to Solve Problems (Part 2)
- 4-17 Logarithmic Functions
- 4-18 Applications of Logarithmic Functions

Unit 5 Transformations of Functions

Translations, Reflections, and Symmetry

5-1 Matching up to Data

5-2 Moving Functions

5-3 More Movement

5-4 Reflecting Functions

5-5 Some Functions Have Symmetry

5-6 Symmetry in Equations

5-7 Expressing Transformations of Functions Algebraically

Scaling Outputs and Inputs

5-8 Scaling the Outputs

5-9 Scaling the Inputs

Putting It All Together

5-10 Combining Functions

5-11 Making a Model for Data

Unit 6 Trigonometric Functions

The Unit Circle

6-1 Moving in Circles

6-2 Revisiting Right Triangles

6-3 The Unit Circle (Part 1)

6-4 The Unit Circle (Part 2)

6-5 The Pythagorean Identity (Part 1)

6-6 The Pythagorean Identity (Part 2)

6-7 Finding Unknown Coordinates on a Circle

Periodic Functions

6-8 Rising and Falling

6-9 Introduction to Trigonometric Functions

6-10 Beyond 2π

6-11 Extending the Domain of Trigonometric Functions

6-12 Tangent

Trigonometry Transformations

6-13 Amplitude and Midline

6-14 Transforming Trigonometric Functions

6-15 Features of Trigonometric Graphs (Part 1)

6-16 Features of Trigonometric Graphs (Part 2)

6-17 Comparing Transformations

6-18 Modeling Circular Motion

Putting It All Together

6-19 Beyond Circles

Unit 7 Statistical Inferences

Study Types
- 7-1 Being Skeptical
- 7-2 Study Types
- 7-3 Randomness in Groups

Distributions
- 7-4 Describing Distributions
- 7-5 Normal Distributions
- 7-6 Areas in Histograms
- 7-7 Areas under a Normal Curve

Not All Samples Are the Same
- 7-8 Not Always Ideal
- 7-9 Variability in Samples
- 7-10 Estimating Proportions from Samples
- 7-11 Reducing Margin of Error
- 7-12 Estimating a Population Mean

Analyzing Experimental Data
- 7-13 Experimenting
- 7-14 Using Normal Distributions for Experiment Analysis
- 7-15 Questioning Experimenting
- 7-16 Heart Rates

Correlation to the Common Core State Standards for Mathematics, Algebra 1

This correlation shows the alignment of *Illustrative Mathematics*, Algebra 1, to the Standards for Mathematical Content from the Common Core State Standards for Mathematics.

* Indicates Mathematical Modeling Standard

Standards for Mathematical Content			Lesson(s)
HSN Number and Quantity			
HSN.RN The Real Number System			
HSN.RN.B	Use properties of rational and irrational numbers.		7-21
HSN.RN.B.3	Explain why the sum or product of two rational numbers is rational; that the sum of a rational number and an irrational number is irrational; and that the product of a nonzero rational number and an irrational number is irrational.		7-20, 7-21
HSN.Q Quantities*			
HSN.Q.A Reason quantitatively and use units to solve problems.			
HSN.Q.A.1	Use units as a way to understand problems and to guide the solution of multi-step problems; choose and interpret units consistently in formulas; choose and interpret the scale and the origin in graphs and data displays.		5-7, 5-8, 5-11
HSN.Q.A.2	Define appropriate quantities for the purpose of descriptive modeling.		2-1, 2-26, 5-17
HSN.Q.A.3	Choose a level of accuracy appropriate to limitations on measurement when reporting quantities.		3-6, 5-11, 5-21
HSA Algebra			
HSA.SSE Seeing Structure in Expressions			
HSA.SSE.A Interpret the structure of expressions.			
HSA.SSE.A.1	Interpret expressions that represent a quantity in terms of its context.*		2-6, 5-4, 5-7, 5-17, 6-2, 6-3
	HSA.SSE.A.1.B Interpret complicated expressions by viewing one or more of their parts as a single entity. *For example, interpret $P(1+r)^n$ as the product of P and a factor not depending on P.*		5-18
HSA.SSE.A.2	Use the structure of an expression to identify ways to rewrite it. *For example, see $x^4 - y^4$ as $(x^2)^2 - (y^2)^2$, thus recognizing it as a difference of squares that can be factored as $(x^2 - y^2)(x^2 + y^2)$.*		6-8, 6-9, 7-6, 7-7, 7-8, 7-10, 7-11, 7-12, 7-14, 7-19, 7-22
HSA.SSE.B Write expressions in equivalent forms to solve problems.			
HSA.SSE.B.3	Choose and produce an equivalent form of an expression to reveal and explain properties of the quantity represented by the expression.*		6-2, 6-8, 6-9, 6-10, 6-13, 7-22
	HSA.SSE.B.3.A Factor a quadratic expression to reveal the zeros of the function it defines.		7-9
	HSA.SSE.B.3.B Complete the square in a quadratic expression to reveal the maximum or minimum value of the function it defines		7-22, 7-23
	HSA.SSE.B.3.C Use the properties of exponents to transform expressions for exponential functions. *For example the expression 1.15^t can be rewritten as $(1.15^{1/12})^{12t} \approx 1.012^{12t}$ to reveal the approximate equivalent monthly interest rate if the annual rate is 15%.*		5-18

Standards for Mathematical Content		Lesson(s)
HSF Functions		
HSF.IF Interpreting Functions		
HSF.IF.A Understand the concept of a function and use function notation.		
HSF.IF.A.1	Understand that a function from one set (called the domain) to another set (called the range) assigns to each element of the domain exactly one element of the range. If f is a function and x is an element of its domain, then $f(x)$ denotes the output of f corresponding to the input x. The graph of f is the graph of the equation $y = f(x)$.	4-1, 4-2, 4-4
HSF.IF.A.2	Use function notation, evaluate functions for inputs in their domains, and interpret statements that use function notation in terms of a context.	4-2, 4-3, 4-4, 4-5, 4-12, 4-17, 5-8, 5-9, 5-11, 5-17, 5-18, 5-19, 6-3, 6-5, 6-14, 7-18
HSF.IF.B Interpret functions that arise in applications in terms of the context.		4-10, 5-8
HSF.IF.B.4	For a function that models a relationship between two quantities, interpret key features of graphs and tables in terms of the quantities, and sketch graphs showing key features given a verbal description of the relationship. *Key features include: intercepts; intervals where the function is increasing, decreasing, positive, or negative; relative maximums and minimums; symmetries; end behavior; and periodicity.**	4-1, 4-2, 4-3, 4-4, 4-5, 4-6, 4-8, 4-9, 4-11, 4-17, 5-1, 5-2, 5-5, 5-6, 5-11, 5-12, 5-13, 5-19, 6-14, 7-10
HSF.IF.B.5	Relate the domain of a function to its graph and, where applicable, to the quantitative relationship it describes. *For example, if the function h(n) gives the number of person-hours it takes to assemble n engines in a factory, then the positive integers would be an appropriate domain for the function.**	4-10, 4-11, 4-12, 5-8, 5-9, 5-11, 5-19, 6-6, 6-7, 7-17
HSF.IF.B.6	Calculate and interpret the average rate of change of a function (presented symbolically or as a table) over a specified interval. Estimate the rate of change from a graph.*	4-7, 4-8, 4-9, 4-18, 5-10, 5-15
HSF.IF.C Analyze functions using different representations.		4-4, 4-12, 4-13, 4-14, 6-4, 6-6, 6-12, 6-15, 6-16, 6-17, 7-22, 7-23
HSF.IF.C.7	Graph functions expressed symbolically and show key features of the graph, by hand in simple cases and using technology for more complicated cases.*	4-12, 5-8, 6-12, 6-13
	HSF.IF.C.7.A Graph linear and quadratic functions and show intercepts, maxima, and minima.	6-6, 6-7, 6-11, 6-13, 6-14, 6-15, 6-16, 6-17, 7-20
	HSF.IF.C.7.B Graph square root, cube root, and piecewise-defined functions, including step functions and absolute value functions.	4-12, 4-13, 4-14
	HSF.IF.C.7.E Graph exponential and logarithmic functions, showing intercepts and end behavior, and trigonometric functions, showing period, midline, and amplitude.	5-9, 5-15
HSF.IF.C.8	Write a function defined by an expression in different but equivalent forms to reveal and explain different properties of the function.	5-18, 6-14
	HSF.IF.C.8.A Use the process of factoring and completing the square in a quadratic function to show zeros, extreme values, and symmetry of the graph, and interpret these in terms of a context.	7-24
	HSF.IF.C.8.B Use the properties of exponents to interpret expressions for exponential functions. *For example, identify percent rate of change in functions such as $y = (1.02)^t$, $y = (0.97)^t$, $y = (1.01)12^t$, $y = (1.2)^t/10$, and classify them as representing exponential growth or decay.*	5-18
HSF.IF.C.9	Compare properties of two functions each represented in a different way (algebraically, graphically, numerically in tables, or by verbal descriptions). *For example, given a graph of one quadratic function and an algebraic expression for another, say which has the larger maximum.*	6-14, 7-23

Standards for Mathematical Content			Lesson(s)
HSF.BF Building Functions			
HSF.BF.A Build a function that models a relationship between two quantities.			
HSF.BF.A.1	Write a function that describes a relationship between two quantities.*		4-17, 4-18, 5-11, 5-15, 6-6
	HSF.BF.A.1.A Determine an explicit expression, a recursive process, or steps for calculation from a context.		4-4, 4-14, 5-2, 5-3, 5-4, 5-15, 5-16, 5-17, 6-1, 6-2, 6-3, 6-4, 6-5, 6-6, 6-7
HSF.BF.B Build new functions from existing functions.			
HSF.BF.B.3	Identify the effect on the graph of replacing $f(x)$ by $f(x) + k$, $k\,f(x)$, $f(kx)$, and $f(x + k)$ for specific values of k (both positive and negative); find the value of k given the graphs. Experiment with cases and illustrate an explanation of the effects on the graph using technology. Include recognizing even and odd functions from their graphs and algebraic expressions for them.		4-14, 6-12, 6-13, 6-15, 6-17
HSF.BF.B.4	Find inverse functions.		4-15, 4-16, 4-17
	HSF.BF.B.4.A Solve an equation of the form $f(x) = c$ for a simple function f that has an inverse and write an expression for the inverse. *For example, $f(x) = 2x^3$ or $f(x) = (x + 1)/(x − 1)$ for $x \neq 1$.*		4-17
HSF.LE Linear, Quadratic, & Exponential Models*			
HSF.LE.A Construct and compare linear, quadratic, and exponential models and solve problems.			
HSF.LE.A.1	Distinguish between situations that can be modeled with linear functions and with exponential functions.		5-11, 5-19, 5-21
	HSF.LE.A.1.A Prove that linear functions grow by equal differences over equal intervals, and that exponential functions grow by equal factors over equal intervals.		5-20
	HSF.LE.A.1.B Recognize situations in which one quantity changes at a constant rate per unit interval relative to another.		5-20, 5-21
	HSF.LE.A.1.C Recognize situations in which a quantity grows or decays by a constant percent rate per unit interval relative to another.		5-11, 5-21
HSF.LE.A.2	Construct linear and exponential functions, including arithmetic and geometric sequences, given a graph, a description of a relationship, or two input-output pairs (include reading these from a table).		5-8, 5-9, 5-11, 5-13, 5-15, 5-19, 5-20, 5-21, 6-12
HSF.LE.A.3	Observe using graphs and tables that a quantity increasing exponentially eventually exceeds a quantity increasing linearly, quadratically, or (more generally) as a polynomial function.		5-1, 5-19, 6-4
HSF.LE.B Interpret expressions for functions in terms of the situation they model.			
HSF.LE.B.5	Interpret the parameters in a linear or exponential function in terms of a context.		5-3, 5-4, 5-5, 5-11, 5-12, 5-13
HSS Statistics & Probability			
HSS.ID Interpreting Categorical & Quantitative Data			
HSS.ID.A Summarize, represent, and interpret data on a single count or measurement variable.			1-16
HSS.ID.A.1	Represent data with plots on the real number line (dot plots, histograms, and box plots)-		1-2, 1-3, 1-4, 1-9, 1-10, 1-14, 1-15, 1-16
HSS.ID.A.2	Use statistics appropriate to the shape of the data distribution to compare center (median, mean) and spread (interquartile range, standard deviation) of two or more different data sets-		1-9, 1-10, 1-11, 1-12, 1-13, 1-14, 1-15, 1-16
HSS.ID.A.3	Interpret differences in shape, center, and spread in the context of the data sets, accounting for possible effects of extreme data points (outliers)-		1-10, 1-12, 1-14, 1-16

Standards for Mathematical Content			Lesson(s)
HSS.ID.B Summarize, represent, and interpret data on two categorical and quantitative variables.			
HSS.ID.B.5	Summarize categorical data for two categories in two-way frequency tables. Interpret relative frequencies in the context of the data (including joint, marginal, and conditional relative frequencies). Recognize possible associations and trends in the data.		3-1, 3-2, 3-3
HSS.ID.B.6	Represent data on two quantitative variables on a scatter plot, and describe how the variables are related.		3-4, 3-7, 3-8, 3-9, 3-10
	HSS.ID.B.6.A	Fit a function to the data; use functions fitted to data to solve problems in the context of the data. Use given functions or choose a function suggested by the context. Emphasize linear, quadratic, and exponential models.	3-4, 3-6, 3-8, 4-17, 4-18, 5-11, 5-21
	HSS.ID.B.6.B	Informally assess the fit of a function by plotting and analyzing residuals.	3-6
	HSS.ID.B.6.C	Fit a linear function for a scatter plot that suggests a linear association.	3-5, 3-6, 4-17, 4-18
HSS.ID.C Interpret linear models.			
HSS.ID.C.7	Interpret the slope (rate of change) and the intercept (constant term) of a linear model in the context of the data.		3-4, 3-5, 3-8, 3-10
HSS.ID.C.8	Compute (using technology) and interpret the correlation coefficient of a linear fit.		3-7, 3-8, 3-10
HSS.ID.C.9	Distinguish between correlation and causation.		3-9, 3-10

Correlation to the Common Core State Standards for Mathematics, Algebra 1 Extra Support Materials

This correlation shows the alignment of *Illustrative Mathematics*, Algebra 1 Extra Support Materials, to the Standards for Mathematical Content from the Common Core State Standards for Mathematics.

* Indicates Mathematical Modeling Standard

Standards for Mathematical Content			Lesson(s)
HSN Number and Quantity			
HSN.RN The Real Number System			
HSN.RN.B Use properties of rational and irrational numbers.			7-15
HSN.RN.B.3		Explain why the sum or product of two rational numbers is rational; that the sum of a rational number and an irrational number is irrational; and that the product of a nonzero rational number and an irrational number is irrational.	7-20, 7-21
HSA Algebra			
HSA.SSE Seeing Structure in Expressions			
HSA.SSE.A Interpret the structure of expressions.			7-16
HSA.SSE.A.1		Interpret expressions that represent a quantity in terms of its context.*	2-1, 2-2, 7-2
HSA.SSE.A.2		Use the structure of an expression to identify ways to rewrite it. *For example, see $x^4 - y^4$ as $(x^2)^2 - (y^2)^2$, thus recognizing it as a difference of squares that can be factored as $(x^2 - y^2)(x^2 + y^2)$.*	2-9, 5-1, 6-3, 6-9, 7-6, 7-7, 7-8, 7-11, 7-12
HSA.SSE.B Write expressions in equivalent forms to solve problems.			
HSA.SSE.B.3		Choose and produce an equivalent form of an expression to reveal and explain properties of the quantity represented by the expression.*	6-8, 6-9, 6-10, 7-4
	HSA.SSE.B.3.A	Factor a quadratic expression to reveal the zeros of the function it defines.	7-10
	HSA.SSE.B.3.C	Use the properties of exponents to transform expressions for exponential functions. *For example the expression 1.15^t can be rewritten as $(1.15^{1/12})^{12t} \approx 1.012^{12t}$ to reveal the approximate equivalent monthly interest rate if the annual rate is 15%.*	5-1
HSA.CED Creating Equations*			
HSA.CED.A Create equations that describe numbers or relationships.			2-1, 2-2
HSA.CED.A.1		Create equations and inequalities in one variable and use them to solve problems. *Include equations arising from linear and quadratic functions, and simple rational and exponential functions.*	2-1, 2-2, 7-1
HSA.CED.A.2		Create equations in two or more variables to represent relationships between quantities; graph equations on coordinate axes with labels and scales.	2-3, 2-4, 2-5, 2-9, 4-16, 4-17
HSA.CED.A.3		Represent constraints by equations or inequalities, and by systems of equations and/or inequalities, and interpret solutions as viable or nonviable options in a modeling context. *For example, represent inequalities describing nutritional and cost constraints on combinations of different foods.*	2-5, 2-12, 2-23, 2-25, 7-2
HSA.CED.A.4		Rearrange formulas to highlight a quantity of interest, using the same reasoning as in solving equations. *For example, rearrange Ohm's law $V = IR$ to highlight resistance R.*	2-3, 2-8, 2-13, 4-16
HSA.REI Reasoning with Equations & Inequalities			
HSA.REI.A Understand solving equations as a process of reasoning and explain the reasoning.			2-6, 2-15, 7-13, 7-17
HSA.REI.A.1		Explain each step in solving a simple equation as following from the equality of numbers asserted at the previous step, starting from the assumption that the original equation has a solution. Construct a viable argument to justify a solution method.	2-7, 2-8, 2-9, 7-5

Standards for Mathematical Content		Lesson(s)
HSA.REI.A.2	Solve simple rational and radical equations in one variable, and give examples showing how extraneous solutions may arise.	2-4
HSA.REI.B Solve equations and inequalities in one variable.		2-6, 2-7
HSA.REI.B.3	Solve linear equations and inequalities in one variable, including equations with coefficients represented by letters.	2-5, 2-18, 2-19, 2-21, 2-26
HSA.REI.B.4	Solve quadratic equations in one variable.	7-2, 7-4, 7-5
	HSA.REI.B.4.A Use the method of completing the square to transform any quadratic equation in x into an equation of the form $(x - p)^2 = q$ that has the same solutions. Derive the quadratic formula from this form.	7-14, 7-15, 7-19
	HSA.REI.B.4.B Solve quadratic equations by inspection (e.g., for $x^2 = 49$), taking square roots, completing the square, the quadratic formula and factoring, as appropriate to the initial form of the equation. Recognize when the quadratic formula gives complex solutions and write them as $a \pm bi$ for real numbers a and b.	7-3, 7-9, 7-12, 7-13, 7-16, 7-18, 7-24
HSA.REI.C Solve systems of equations.		2-7, 2-16
HSA.REI.C.5	Prove that, given a system of two equations in two variables, replacing one equation by the sum of that equation and a multiple of the other produces a system with the same solutions.	2-14, 2-15, 2-16
HSA.REI.C.6	Solve systems of linear equations exactly and approximately (e.g., with graphs), focusing on pairs of linear equations in two variables.	2-6, 2-12, 2-13, 2-14, 2-15, 2-16, 2-17
HSA.REI.C.7	Solve a simple system consisting of a linear equation and a quadratic equation in two variables algebraically and graphically. For example, find the points of intersection between the line $y = -3x$ and the circle $x^2 + y^2 = 3$.	7-24
HSA.REI.D Represent and solve equations and inequalities graphically.		2-25, 7-10
HSA.REI.D.10	Understand that the graph of an equation in two variables is the set of all its solutions plotted in the coordinate plane, often forming a curve (which could be a line).	2-5, 2-21, 2-22
HSA.REI.D.11	Explain why the x-coordinates of the points where the graphs of the equations $y = f(x)$ and $y = g(x)$ intersect are the solutions of the equation $f(x) = g(x)$; find the solutions approximately, e.g., using technology to graph the functions, make tables of values, or find successive approximations. Include cases where $f(x)$ and/or $g(x)$ are linear, polynomial, rational, absolute value, exponential, and logarithmic functions.*	2-10, 2-11
HSA.REI.D.12	Graph the solutions to a linear inequality in two variables as a half-plane (excluding the boundary in the case of a strict inequality), and graph the solution set to a system of linear inequalities in two variables as the intersection of the corresponding half-planes.	2-10, 2-11, 2-21, 2-22, 2-23, 2-24

HSF Functions

HSF.IF Interpreting Functions

HSF.IF.A Understand the concept of a function and use function notation.		
HSF.IF.A.1	Understand that a function from one set (called the domain) to another set (called the range) assigns to each element of the domain exactly one element of the range. If f is a function and x is an element of its domain, then $f(x)$ denotes the output of f corresponding to the input x. The graph of f is the graph of the equation $y = f(x)$.	5-8, 5-10
HSF.IF.A.2	Use function notation, evaluate functions for inputs in their domains, and interpret statements that use function notation in terms of a context.	4-3, 5-8, 5-10, 5-12, 5-20, 6-4, 6-5, 6-11, 6-12
HSF.IF.B Interpret functions that arise in applications in terms of the context.		
HSF.IF.B.4	For a function that models a relationship between two quantities, interpret key features of graphs and tables in terms of the quantities, and sketch graphs showing key features given a verbal description of the relationship. Key features include: intercepts; intervals where the function is increasing, decreasing, positive, or negative; relative maximums and minimums; symmetries; end behavior; and periodicity.*	4-1, 4-2, 4-3, 4-6, 4-8, 4-9, 4-11, 5-2, 5-12, 6-10, 6-11

	Standards for Mathematical Content	Lesson(s)
HSF.IF.B.5	Relate the domain of a function to its graph and, where applicable, to the quantitative relationship it describes. *For example, if the function h(n) gives the number of person-hours it takes to assemble n engines in a factory, then the positive integers would be an appropriate domain for the function.**	4-10, 4-11, 5-8, 5-9, 5-19, 6-7
HSF.IF.B.6	Calculate and interpret the average rate of change of a function (presented symbolically or as a table) over a specified interval. Estimate the rate of change from a graph.*	4-7, 4-18, 5-10
HSF.IF.C Analyze functions using different representations.		4-4, 4-5, 6-11, 7-22, 7-23, 7-24
HSF.IF.C.7	Graph functions expressed symbolically and show key features of the graph, by hand in simple cases and using technology for more complicated cases.*	6-11
	HSF.IF.C.7.A Graph linear and quadratic functions and show intercepts, maxima, and minima.	6-5, 6-6, 6-7, 6-10, 6-12, 6-15, 6-16
	HSF.IF.C.7.B Graph square root, cube root, and piecewise-defined functions, including step functions and absolute value functions.	4-12, 4-13, 4-14
	HSF.IF.C.7.E Graph exponential and logarithmic functions, showing intercepts and end behavior, and trigonometric functions, showing period, midline, and amplitude.	5-13
HSF.IF.C.8	Write a function defined by an expression in different but equivalent forms to reveal and explain different properties of the function.	6-11
	HSF.IF.C.8.A Use the process of factoring and completing the square in a quadratic function to show zeros, extreme values, and symmetry of the graph, and interpret these in terms of a context.	6-15
	HSF.IF.C.8.B Use the properties of exponents to interpret expressions for exponential functions. *For example, identify percent rate of change in functions such as $y = (1.02)^t$, $y = (0.97)^t$, $y = (1.01)12^t$, $y = (1.2)^t /10$, and classify them as representing exponential growth or decay.*	5-18
HSF.IF.C.9	Compare properties of two functions each represented in a different way (algebraically, graphically, numerically in tables, or by verbal descriptions). *For example, given a graph of one quadratic function and an algebraic expression for another, say which has the larger maximum.*	6-14
HSF.BF Building Functions		
HSF.BF.A Build a function that models a relationship between two quantities.		
HSF.BF.A.1	Write a function that describes a relationship between two quantities.*	5-11
	HSF.BF.A.1.A Determine an explicit expression, a recursive process, or steps for calculation from a context.	5-15, 5-17, 6-1, 6-2, 6-3, 6-5
HSF.BF.B Build new functions from existing functions.		
HSF.BF.B.3	Identify the effect on the graph of replacing $f(x)$ by $f(x) + k$, $k f(x)$, $f(kx)$, and $f(x + k)$ for specific values of k (both positive and negative); find the value of k given the graphs. Experiment with cases and illustrate an explanation of the effects on the graph using technology. Include recognizing even and odd functions from their graphs and algebraic expressions for them.	6-12, 6-15, 6-17
HSF.BF.B.4	Find inverse functions.	4-15, 5-11
HSF.LE Linear, Quadratic, & Exponential Models*		
HSF.LE.A Construct and compare linear, quadratic, and exponential models and solve problems.		6-1
HSF.LE.A.1	Distinguish between situations that can be modeled with linear functions and with exponential functions.	5-2, 5-11, 5-21, 6-2
	HSF.LE.A.1.A Prove that linear functions grow by equal differences over equal intervals, and that exponential functions grow by equal factors over equal intervals.	5-20

Standards for Mathematical Content		Lesson(s)
HSF.LE.A.2	Construct linear and exponential functions, including arithmetic and geometric sequences, given a graph, a description of a relationship, or two input-output pairs (include reading these from a table).	5-5, 5-6, 5-11, 5-12, 5-13, 5-15, 5-16, 5-21, 6-6
HSF.LE.A.3	Observe using graphs and tables that a quantity increasing exponentially eventually exceeds a quantity increasing linearly, quadratically, or (more generally) as a polynomial function.	5-1, 5-19, 6-4
HSF.LE.B Interpret expressions for functions in terms of the situation they model.		
HSF.LE.B.5	Interpret the parameters in a linear or exponential function in terms of a context.	5-1, 5-3, 5-12
HSS Statistics & Probability		
HSS.ID Interpreting Categorical & Quantitative Data		
HSS.ID.A Summarize, represent, and interpret data on a single count or measurement variable.		1-13
HSS.ID.A.1	Represent data with plots on the real number line (dot plots, histograms, and box plots)-	1-1, 1-2, 1-3, 1-4, 1-9, 1-16
HSS.ID.A.2	Use statistics appropriate to the shape of the data distribution to compare center (median, mean) and spread (interquartile range, standard deviation) of two or more different data sets-	1-1, 1-3, 1-4, 1-5, 1-9, 1-11, 1-12, 1-13, 1-15, 1-16
HSS.ID.A.3	Interpret differences in shape, center, and spread in the context of the data sets, accounting for possible effects of extreme data points (outliers)-	1-5, 1-10, 1-14, 1-16
HSS.ID.B Summarize, represent, and interpret data on two categorical and quantitative variables.		3-7, 3-9, 3-10
HSS.ID.B.6	Represent data on two quantitative variables on a scatter plot, and describe how the variables are related.	3-8
	HSS.ID.B.6.A Fit a function to the data; use functions fitted to data to solve problems in the context of the data. Use given functions or choose a function suggested by the context. Emphasize linear, quadratic, and exponential models.	4-18, 5-11, 5-21
HSS.ID.C Interpret linear models.		3-7, 3-9, 3-10
HSS.ID.C.7	Interpret the slope (rate of change) and the intercept (constant term) of a linear model in the context of the data.	3-7
HSS.ID.C.8	Compute (using technology) and interpret the correlation coefficient of a linear fit.	3-7, 3-8
HSS.ID.C.9	Distinguish between correlation and causation.	3-9

Correlation to the Common Core State Standards for Mathematics, Geometry

This correlation shows the alignment of *Illustrative Mathematics*, Geometry, to the Standards for Mathematical Content from the Common Core State Standards for Mathematics.

* Indicates Mathematical Modeling Standard

Standards for Mathematical Content		Lesson(s)
HSN Number and Quantity		
HSN.Q Quantities*		
HSN.Q.A Reason quantitatively and use units to solve problems.		
HSN.Q.A.1	Use units as a way to understand problems and to guide the solution of multi-step problems; choose and interpret units consistently in formulas; choose and interpret the scale and the origin in graphs and data displays.	3-2, 3-12, 5-6, 5-7, 5-8, 5-17, 6-10, 6-11, 6-16
HSN.Q.A.2	Define appropriate quantities for the purpose of descriptive modeling.	1-9, 4-1, 4-7, 4-10, 4-11
HSN.Q.A.3	Choose a level of accuracy appropriate to limitations on measurement when reporting quantities.	1-9, 1-13, 1-14, 2-4, 3-16, 4-2, 4-4
HSA Algebra		
HSA.SSE Seeing Structure in Expressions		
HSA.SSE.A Interpret the structure of expressions.		6-8, 6-9
HSA.SSE.A.1	Interpret expressions that represent a quantity in terms of its context.*	7-13
	HSA.SSE.A.1.A Interpret parts of an expression, such as terms, factors, and coefficients.	5-4, 5-6, 6-9, 6-15
	HSA.SSE.A.1.B Interpret complicated expressions by viewing one or more of their parts as a single entity. For example, interpret $P(1+r)^n$ as the product of P and a factor not depending on P.	5-9, 7-8, 7-11
HSA.SSE.A.2	Use the structure of an expression to identify ways to rewrite it. For example, see $x^4 - y^4$ as $(x^2)^2 - (y^2)^2$, thus recognizing it as a difference of squares that can be factored as $(x^2 - y^2)(x^2 + y^2)$.	6-5
HSA.SSE.B Write expressions in equivalent forms to solve problems.		
HSA.SSE.B.3	Choose and produce an equivalent form of an expression to reveal and explain properties of the quantity represented by the expression.*	6-5, 6-6, 6-8
HSA.CED Creating Equations*		
HSA.CED.A Create equations that describe numbers or relationships.		
HSA.CED.A.2	Create equations in two or more variables to represent relationships between quantities; graph equations on coordinate axes with labels and scales.	5-5, 5-7, 6-12, 7-13
HSA.CED.A.4	Rearrange formulas to highlight a quantity of interest, using the same reasoning as in solving equations. For example, rearrange Ohm's law $V = IR$ to highlight resistance R.	3-14, 6-12
HSA.REI Reasoning with Equations & Inequalities		
HSA.REI.C Solve systems of equations.		
HSA.REI.C.7	Solve a simple system consisting of a linear equation and a quadratic equation in two variables algebraically and graphically. For example, find the points of intersection between the line $y = -3x$ and the circle $x^2 + y^2 = 3$.	6-13
HSF Functions		
HSF.IF Interpreting Functions		
HSF.IF.C Analyze functions using different representations.		
HSF.IF.C.7	HSF.IF.C.7.B Graph square root, cube root, and piecewise-defined functions, including step functions and absolute value functions.	5-5, 5-7, 5-18

Standards for Mathematical Content		Lesson(s)
HSG Geometry		
HSG.CO Congruence		
HSG.CO.A Experiment with transformations in the plane.		
HSG.CO.A.1	Know precise definitions of angle, circle, perpendicular line, parallel line, and line segment, based on the undefined notions of point, line, distance along a line, and distance around a circular arc.	1-1, 1-2, 1-3, 1-6, 1-8, 1-13, 1-20, 2-4, 6-4
HSG.CO.A.2	Represent transformations in the plane using, e.g., transparencies and geometry software; describe transformations as functions that take points in the plane as inputs and give other points as outputs. Compare transformations that preserve distance and angle to those that do not (e.g., translation versus horizontal stretch).	1-10, 1-11, 1-13, 1-15, 1-17, 1-18, 1-20, 1-21, 3-1, 3-3, 6-1, 6-2, 6-3, 6-11, 6-15
HSG.CO.A.3	Given a rectangle, parallelogram, trapezoid, or regular polygon, describe the rotations and reflections that carry it onto itself.	1-15, 1-16
HSG.CO.A.4	Develop definitions of rotations, reflections, and translations in terms of angles, circles, perpendicular lines, parallel lines, and line segments.	1-11, 1-12, 1-14
HSG.CO.A.5	Given a geometric figure and a rotation, reflection, or translation, draw the transformed figure using, e.g., graph paper, tracing paper, or geometry software. Specify a sequence of transformations that will carry a given figure onto another.	1-10, 1-13, 1-17, 1-18, 2-1, 6-1, 6-2, 6-3
HSG.CO.B Understand congruence in terms of rigid motions.		6-1
HSG.CO.B.6	Use geometric descriptions of rigid motions to transform figures and to predict the effect of a given rigid motion on a given figure; given two figures, use the definition of congruence in terms of rigid motions to decide if they are congruent.	2-1, 2-2, 2-5
HSG.CO.B.7	Use the definition of congruence in terms of rigid motions to show that two triangles are congruent if and only if corresponding pairs of sides and corresponding pairs of angles are congruent.	2-3
HSG.CO.B.8	Explain how the criteria for triangle congruence (ASA, SAS, and SSS) follow from the definition of congruence in terms of rigid motions.	2-4, 2-6, 2-7, 2-9
HSG.CO.C Prove geometric theorems.		
HSG.CO.C.9	Prove theorems about lines and angles. *Theorems include: vertical angles are congruent; when a transversal crosses parallel lines, alternate interior angles are congruent and corresponding angles are congruent; points on a perpendicular bisector of a line segment are exactly those equidistant from the segment's endpoints.*	1-19, 1-20, 1-21, 2-8, 2-14, 7-6
HSG.CO.C.10	Prove theorems about triangles. *Theorems include: measures of interior angles of a triangle sum to 180°; base angles of isosceles triangles are congruent; the segment joining midpoints of two sides of a triangle is parallel to the third side and half the length; the medians of a triangle meet at a point.*	1-21, 2-6, 2-7, 3-5, 6-16, 6-17, 7-5, 7-6, 7-7
HSG.CO.C.11	Prove theorems about parallelograms. *Theorems include: opposite sides are congruent, opposite angles are congruent, the diagonals of a parallelogram bisect each other, and conversely, rectangles are parallelograms with congruent diagonals.*	2-7, 2-9, 2-10, 2-12, 2-13, 2-15
HSG.CO.D Make geometric constructions.		
HSG.CO.D.12	Make formal geometric constructions with a variety of tools and methods (compass and straightedge, string, reflective devices, paper folding, dynamic geometric software, etc.). *Copying a segment; copying an angle; bisecting a segment; bisecting an angle; constructing perpendicular lines, including the perpendicular bisector of a line segment; and constructing a line parallel to a given line through a point not on the line.*	1-1, 1-2, 1-3, 1-4, 1-5, 1-6, 1-8, 1-9, 1-22
HSG.CO.D.13	Construct an equilateral triangle, a square, and a regular hexagon inscribed in a circle.	1-2, 1-4, 1-7, 1-8, 1-22
HSG.SRT Similarity, Right Triangles, & Trigonometry		
HSG.SRT.A Understand similarity in terms of similarity transformations.		
HSG.SRT.A.1	Verify experimentally the properties of dilations given by a center and a scale factor:	3-1, 3-3, 3-4
	HSG.SRT.A.1.A A dilation takes a line not passing through the center of the dilation to a parallel line, and leaves a line passing through the center unchanged.	3-4
	HSG.SRT.A.1.B The dilation of a line segment is longer or shorter in the ratio given by the scale factor.	3-1, 3-3, 3-6
HSG.SRT.A.2	Given two figures, use the definition of similarity in terms of similarity transformations to decide if they are similar; explain using similarity transformations the meaning of similarity for triangles as the equality of all corresponding pairs of angles and the proportionality of all corresponding pairs of sides.	3-6, 3-7, 3-8
HSG.SRT.A.3	Use the properties of similarity transformations to establish the AA criterion for two triangles to be similar.	3-9

Standards for Mathematical Content		Lesson(s)
HSG.SRT.B Prove theorems involving similarity.		
HSG.SRT.B.4	Prove theorems about triangles. *Theorems include: a line parallel to one side of a triangle divides the other two proportionally, and conversely; the Pythagorean Theorem proved using triangle similarity.*	3-5, 3-11, 3-14
HSG.SRT.B.5	Use congruence and similarity criteria for triangles to solve problems and to prove relationships in geometric figures.	3-6, 3-11, 3-12, 3-13, 3-15, 3-16, 4-1, 4-2, 4-3, 6-3, 7-1, 7-2, 7-6
HSG.SRT.C Define trigonometric ratios and solve problems involving right triangles.		4-7, 4-10, 4-11
HSG.SRT.C.6	Understand that by similarity, side ratios in right triangles are properties of the angles in the triangle, leading to definitions of trigonometric ratios for acute angles.	4-4, 4-5, 4-6, 4-9
HSG.SRT.C.7	Explain and use the relationship between the sine and cosine of complementary angles.	4-8
HSG.SRT.C.8	Use trigonometric ratios and the Pythagorean Theorem to solve right triangles in applied problems.*	4-7, 4-9, 4-10, 5-11, 7-14
HSG.C Circles		
HSG.C.A Understand and apply theorems about circles.		
HSG.C.A.1	Prove that all circles are similar.	3-8
HSG.C.A.2	Identify and describe relationships among inscribed angles, radii, and chords. *Include the relationship between central, inscribed, and circumscribed angles; inscribed angles on a diameter are right angles; the radius of a circle is perpendicular to the tangent where the radius intersects the circle.*	6-14, 7-1, 7-2, 7-3, 7-14
HSG.C.A.3	Construct the inscribed and circumscribed circles of a triangle, and prove properties of angles for a quadrilateral inscribed in a circle.	7-4, 7-5, 7-7, 7-14
HSG.C.B Find arc lengths and areas of sectors of circles.		7-8, 7-9, 7-14
HSG.C.B.5	Derive using similarity the fact that the length of the arc intercepted by an angle is proportional to the radius, and define the radian measure of the angle as the constant of proportionality; derive the formula for the area of a sector	7-8, 7-11, 7-12, 7-13
HSG.GPE Expressing Geometric Properties with Equations		
HSG.GPE.A Translate between the geometric description and the equation for a conic section.		
HSG.GPE.A.1	Derive the equation of a circle of given center and radius using the Pythagorean Theorem; complete the square to find the center and radius of a circle given by an equation.	6-4, 6-6
HSG.GPE.A.2	Derive the equation of a parabola given a focus and directrix.	6-8
HSG.GPE.B Use coordinates to prove simple geometric theorems algebraically.		
HSG.GPE.B.4	Use coordinates to prove simple geometric theorems algebraically. *For example, prove or disprove that a figure defined by four given points in the coordinate plane is a rectangle; prove or disprove that the point $(1, \sqrt{3})$ lies on the circle centered at the origin and containing the point $(0, 2)$.*	6-4, 6-7, 6-10, 6-14, 6-16, 6-17
HSG.GPE.B.5	Prove the slope criteria for parallel and perpendicular lines and use them to solve geometric problems (e.g., find the equation of a line parallel or perpendicular to a given line that passes through a given point).	6-10, 6-11, 6-12, 6-13, 6-14, 6-17
HSG.GPE.B.6	Find the point on a directed line segment between two given points that partitions the segment in a given ratio.	6-15, 6-16
HSG.GPE.B.7	Use coordinates to compute perimeters of polygons and areas of triangles and rectangles, e.g., using the distance formula.*	6-14
HSG.GMD Geometric Measurement & Dimension		5-4, 5-5, 5-6, 5-7, 5-8
HSG.GMD.A Explain volume formulas and use them to solve problems.		
HSG.GMD.A.1	Give an informal argument for the formulas for the circumference of a circle, area of a circle, volume of a cylinder, pyramid, and cone. *Use dissection arguments, Cavalieri's principle, and informal limit arguments.*	4-11, 5-9, 5-10, 5-13, 7-8, 7-10
HSG.GMD.A.3	Use volume formulas for cylinders, pyramids, cones, and spheres to solve problems.*	5-9, 5-11, 5-14, 5-15, 5-16
HSG.GMD.B Visualize relationships between two-dimensional and three-dimensional objects.		
HSG.GMD.B.4	Identify the shapes of two-dimensional cross-sections of three-dimensional objects, and identify three-dimensional objects generated by rotations of two-dimensional objects.	5-1, 5-2, 5-3, 5-8, 5-9, 5-10, 5-11, 5-15

Standards for Mathematical Content		Lesson(s)
HSG.MG Modeling with Geometry		
HSG.MG.A Apply geometric concepts in modeling situations.		
HSG.MG.A.1	Use geometric shapes, their measures, and their properties to describe objects (e.g., modeling a tree trunk or a human torso as a cylinder).*	5-8, 5-10, 5-16, 5-18
HSG.MG.A.2	Apply concepts of density based on area and volume in modeling situations (e.g., persons per square mile, BTUs per cubic foot).*	5-17
HSG.MG.A.3	Apply geometric methods to solve design problems (e.g., designing an object or structure to satisfy physical constraints or minimize cost; working with typographic grid systems based on ratios).*	1-9, 2-10, 3-2, 4-1, 5-7, 5-8, 5-14, 5-16, 5-18, 7-14
HSS Statistics & Probability		
HSS.ID Interpreting Categorical & Quantitative Data		
HSS.ID.B Summarize, represent, and interpret data on two categorical and quantitative variables.		
HSS.ID.B.5	Summarize categorical data for two categories in two-way frequency tables. Interpret relative frequencies in the context of the data (including joint, marginal, and conditional relative frequencies). Recognize possible associations and trends in the data.	8-4
HSS.CP Conditional Probability & the Rules of Probability		
HSS.CP.A Understand independence and conditional probability and use them to interpret data.		
HSS.CP.A.1	Describe events as subsets of a sample space (the set of outcomes) using characteristics (or categories) of the outcomes, or as unions, intersections, or complements of other events ("or," "and," "not").	8-2, 8-3, 8-5
HSS.CP.A.2	Understand that two events A and B are independent if the probability of A and B occurring together is the product of their probabilities, and use this characterization to determine if they are independent.	8-7, 8-10, 8-11
HSS.CP.A.3	Understand the conditional probability of A given B as $P(A \text{ and } B)/P(B)$, and interpret independence of A and B as saying that the conditional probability of A given B is the same as the probability of A, and the conditional probability of B given A is the same as the probability of B.	8-8, 8-10, 8-11
HSS.CP.A.4	Construct and interpret two-way frequency tables of data when two categories are associated with each object being classified. Use the two-way table as a sample space to decide if events are independent and to approximate conditional probabilities. *For example, collect data from a random sample of students in your school on their favorite subject among math, science, and English. Estimate the probability that a randomly selected student from your school will favor science given that the student is in tenth grade. Do the same for other subjects and compare the results.*	8-4, 8-7, 8-9, 8-10
HSS.CP.A.5	Recognize and explain the concepts of conditional probability and independence in everyday language and everyday situations. *For example, compare the chance of having lung cancer if you are a smoker with the chance of being a smoker if you have lung cancer.*	8-7, 8-10, 8-11
HSS.CP.B Use the rules of probability to compute probabilities of compound events.		
HSS.CP.B.6	Find the conditional probability of A given B as the fraction of B's outcomes that also belong to A, and interpret the answer in terms of the model.	8-8, 8-11
HSS.CP.B.7	Apply the Addition Rule, $P(A \text{ or } B) = P(A) + P(B) - P(A \text{ and } B)$, and interpret the answer in terms of the model.	8-6

Correlation to the Common Core State Standards for Mathematics, Algebra 2

This correlation shows the alignment of *Illustrative Mathematics*, Algebra 2, to the Standards for Mathematical Content from the Common Core State Standards for Mathematics.

* Indicates Mathematical Modeling Standard

Standards for Mathematical Content		Lesson(s)
HSN Number and Quantity		
HSN.RN The Real Number System		
HSN.RN.A Extend the properties of exponents to rational exponents.		
HSN.RN.A.1	Explain how the definition of the meaning of rational exponents follows from extending the properties of integer exponents to those values, allowing for a notation for radicals in terms of rational exponents. *For example, we define $5^{1/3}$ to be the cube root of 5 because we want $(5^{1/3})^3 = 5^{(1/3)3}$ to hold, so $(51/3)3$ must equal 5.*	3-3, 3-4, 3-5, 4-3, 4-6, 4-7
HSN.RN.A.2	Rewrite expressions involving radicals and rational exponents using the properties of exponents.	3-3, 3-4, 3-5
HSN.Q Quantities*		
HSN.Q.A Reason quantitatively and use units to solve problems.		
HSN.Q.A.1	Use units as a way to understand problems and to guide the solution of multi-step problems; choose and interpret units consistently in formulas; choose and interpret the scale and the origin in graphs and data displays.	6-18
HSN.CN The Complex Number System		
HSN.CN.A Perform arithmetic operations with complex numbers.		
HSN.CN.A.1	Know there is a complex number i such that $i^2 = -1$, and every complex number has the form $a + bi$ with a and b real.	3-10, 3-11, 3-12, 3-13, 3-14
HSN.CN.A.2	Use the relation $i^2 = -1$ and the commutative, associative, and distributive properties to add, subtract, and multiply complex numbers.	3-12, 3-13, 3-14, 3-15
HSN.CN.C Use complex numbers in polynomial identities and equations.		
HSN.CN.C.7	Solve quadratic equations with real coefficients that have complex solutions.	3-17, 3-18, 3-19
HSA Algebra		
HSA.SSE Seeing Structure in Expressions		
HSA.SSE.A Interpret the structure of expressions.		2-3, 2-8, 2-9, 4-5
HSA.SSE.A.1	Interpret expressions that represent a quantity in terms of its context.*	2-2, 2-18, 2-19, 2-26, 4-2, 4-4, 4-6, 4-7
	HSA.SSE.A.1.A Interpret parts of an expression, such as terms, factors, and coefficients.	2-1, 2-7
	HSA.SSE.A.1.B Interpret complicated expressions by viewing one or more of their parts as a single entity. *For example, interpret $P(1+r)^n$ as the product of P and a factor not depending on P.*	4-13
HSA.SSE.A.2	Use the structure of an expression to identify ways to rewrite it. *For example, see $x^4 - y^4$ as $(x^2)^2 - (y^2)^2$, thus recognizing it as a difference of squares that can be factored as $(x^2 - y^2)(x^2 + y^2)$.*	2-23, 2-25
HSA.SSE.B Write expressions in equivalent forms to solve problems.		
HSA.SSE.B.3	Choose and produce an equivalent form of an expression to reveal and explain properties of the quantity represented by the expression.*	2-6, 4-7, 4-10
	HSA.SSE.B.3.C Use the properties of exponents to transform expressions for exponential functions. *For example the expression $1 - 15^t$ can be rewritten as $(1 - 15^{1/12})^{12t} \approx 1 - 012^{12t}$ to reveal the approximate equivalent monthly interest rate if the annual rate is 15%.*	4-4

Standards for Mathematical Content			Lesson(s)
HSA.SSE.B.4		Derive the formula for the sum of a finite geometric series (when the common ratio is not 1), and use the formula to solve problems. *For example, calculate mortgage payments.**	2-25, 2-26
HSA.APR Arithmetic with Polynomials & Rational Expressions			
HSA.APR.A Perform arithmetic operations on polynomials.			2-6, 2-12, 2-13, 2-14
HSA.APR.A.1		Understand that polynomials form a system analogous to the integers, namely, they are closed under the operations of addition, subtraction, and multiplication; add, subtract, and multiply polynomials.	2-2, 2-4, 2-6
HSA.APR.B Understand the relationship between zeros and factors of polynomials.			2-5, 2-6, 2-7, 2-14
HSA.APR.B.2		Know and apply the Remainder Theorem: For a polynomial $p(x)$ and a number a, the remainder on division by $x - a$ is $p(a)$, so $p(a) = 0$ if and only if $(x - a)$ is a factor of $p(x)$.	2-15
HSA.APR.B.3		Identify zeros of polynomials when suitable factorizations are available, and use the zeros to construct a rough graph of the function defined by the polynomial.	2-5, 2-10, 2-12, 2-14
HSA.APR.C Use polynomial identities to solve problems.			2-25
HSA.APR.C.4		Prove polynomial identities and use them to describe numerical relationships. For example, the polynomial identity $(x^2 + y^2)^2 = (x^2 - y^2)^2 + (2xy)^2$ can be used to generate Pythagorean triples.	2-23, 2-24
HSA.APR.D Rewrite rational expressions.			2-24
HSA.APR.D.6		Rewrite simple rational expressions in different forms; write $a(x)/b(x)$ in the form $q(x) + r(x)/b(x)$, where $a(x)$, $b(x)$, $q(x)$, and $r(x)$ are polynomials with the degree of $r(x)$ less than the degree of $b(x)$, using inspection, long division, or, for the more complicated examples, a computer algebra system.	2-18, 2-19
HSA.CED Creating Equations*			
HSA.CED.A Create equations that describe numbers or relationships.			
HSA.CED.A.1		Create equations and inequalities in one variable and use them to solve problems. *Include equations arising from linear and quadratic functions, and simple rational and exponential functions.*	2-20, 2-21
HSA.CED.A.2		Create equations in two or more variables to represent relationships between quantities; graph equations on coordinate axes with labels and scales.	2-1, 2-2, 2-16, 2-17, 2-20
HSA.CED.A.4		Rearrange formulas to highlight a quantity of interest, using the same reasoning as in solving equations. *For example, rearrange Ohm's law $V = IR$ to highlight resistance R.*	2-16
HSA.REI Reasoning with Equations & Inequalities			
HSA.REI.A Understand solving equations as a process of reasoning and explain the reasoning.			
HSA.REI.A.1		Explain each step in solving a simple equation as following from the equality of numbers asserted at the previous step, starting from the assumption that the original equation has a solution. Construct a viable argument to justify a solution method.	2-20, 2-21, 3-7
HSA.REI.A.2		Solve simple rational and radical equations in one variable, and give examples showing how extraneous solutions may arise.	2-22, 3-6, 3-7, 3-8, 3-9
HSA.REI.B Solve equations and inequalities in one variable.			
HSA.REI.B.4		Solve quadratic equations in one variable.	3-18
	HSA.REI.B.4.A	Use the method of completing the square to transform any quadratic equation in x into an equation of the form $(x - p)^2 = q$ that has the same solutions. Derive the quadratic formula from this form.	3-16
	HSA.REI.B.4.B	Solve quadratic equations by inspection (e.g., for $x^2 = 49$), taking square roots, completing the square, the quadratic formula and factoring, as appropriate to the initial form of the equation. Recognize when the quadratic formula gives complex solutions and write them as $a \pm bi$ for real numbers a and b.	3-7, 3-16, 3-17, 3-19

Standards for Mathematical Content		Lesson(s)
HSA.REI.C Solve systems of equations.		
HSA.REI.C.7	Solve a simple system consisting of a linear equation and a quadratic equation in two variables algebraically and graphically. For example, find the points of intersection between the line $y = -3x$ and the circle $x^2 + y^2 = 3$.	2-11
HSA.REI.D Represent and solve equations and inequalities graphically.		
HSA.REI.D.11	Explain why the x-coordinates of the points where the graphs of the equations $y = f(x)$ and $y = g(x)$ intersect are the solutions of the equation $f(x) = g(x)$; find the solutions approximately, e.g., using technology to graph the functions, make tables of values, or find successive approximations. Include cases where $f(x)$ and/or $g(x)$ are linear, polynomial, rational, absolute value, exponential, and logarithmic functions.*	2-2, 2-11, 2-21, 3-8, 3-17, 4-15, 4-16
HSF Functions		
HSF.IF Interpreting Functions		
HSF.IF.A Understand the concept of a function and use function notation.		
HSF.IF.A.2	Use function notation, evaluate functions for inputs in their domains, and interpret statements that use function notation in terms of a context.	2-2, 4-12
HSF.IF.A.3	Recognize that sequences are functions, sometimes defined recursively, whose domain is a subset of the integers. For example, the Fibonacci sequence is defined recursively by $f(0) = f(1) = 1, f(n+1) = f(n) + f(n-1)$ for $n \geq 1$.	1-5, 1-7, 1-9
HSF.IF.B Interpret functions that arise in applications in terms of the context.		
HSF.IF.B.4	For a function that models a relationship between two quantities, interpret key features of graphs and tables in terms of the quantities, and sketch graphs showing key features given a verbal description of the relationship. Key features include: intercepts; intervals where the function is increasing, decreasing, positive, or negative; relative maximums and minimums; symmetries; end behavior; and periodicity.*	2-1, 2-17, 4-18, 5-11, 6-8, 6-15, 6-18
HSF.IF.B.5	Relate the domain of a function to its graph and, where applicable, to the quantitative relationship it describes. For example, if the function h(n) gives the number of person-hours it takes to assemble n engines in a factory, then the positive integers would be an appropriate domain for the function.*	1-9, 2-1
HSF.IF.C Analyze functions using different representations.		1-3, 1-4, 1-6, 2-9, 2-17, 2-18, 4-17, 5-5, 5-10, 6-8
HSF.IF.C.7	Graph functions expressed symbolically and show key features of the graph, by hand in simple cases and using technology for more complicated cases.*	2-3, 2-17, 4-13, 6-9, 6-12
	HSF.IF.C.7.C Graph polynomial functions, identifying zeros when suitable factorizations are available, and showing end behavior.	2-10
	HSF.IF.C.7.E Graph exponential and logarithmic functions, showing intercepts and end behavior, and trigonometric functions, showing period, midline, and amplitude.	4-15, 4-17, 6-13, 6-14, 6-15, 6-16, 6-17, 6-18, 6-19
HSF.IF.C.8	Write a function defined by an expression in different but equivalent forms to reveal and explain different properties of the function.	5-6
	HSF.IF.C.8.B Use the properties of exponents to interpret expressions for exponential functions. For example, identify percent rate of change in functions such as $y = (1 - 02)^t, y = (0 - 97)^t, y = (1 - 01)^{12t}, y = (1 - 2)^t /10$, and classify them as representing exponential growth or decay.	4-6

Standards for Mathematical Content			Lesson(s)
HSF.BF Building Functions			
HSF.BF.A Build a function that models a relationship between two quantities.			
HSF.BF.A.1	Write a function that describes a relationship between two quantities.*		5-7
	HSF.BF.A.1.A Determine an explicit expression, a recursive process, or steps for calculation from a context.		1-11, 4-8
	HSF.BF.A.1.B Combine standard function types using arithmetic operations. *For example, build a function that models the temperature of a cooling body by adding a constant function to a decaying exponential, and relate these functions to the model.*		5-10, 5-11
HSF.BF.A.2	Write arithmetic and geometric sequences both recursively and with an explicit formula, use them to model situations, and translate between the two forms.*		1-5, 1-6, 1-7, 1-8, 1-9, 1-10, 1-11
HSF.BF.B Build new functions from existing functions.			
HSF.BF.B.3	Identify the effect on the graph of replacing $f(x)$ by $f(x) + k$, $k f(x)$, $f(kx)$, and $f(x + k)$ for specific values of k (both positive and negative); find the value of k given the graphs. Experiment with cases and illustrate an explanation of the effects on the graph using technology. Include recognizing even and odd functions from their graphs and algebraic expressions for them.		5-1, 5-2, 5-3, 5-4, 5-5, 5-6, 5-7, 5-8, 5-9, 5-11, 6-15, 6-17
HSF.LE Linear, Quadratic, & Exponential Models*			
HSF.LE.A Construct and compare linear, quadratic, and exponential models and solve problems.			4-13
HSF.LE.A.1	**HSF.LE.A.1.A** Prove that linear functions grow by equal differences over equal intervals, and that exponential functions grow by equal factors over equal intervals.		4-5
	HSF.LE.A.1.B Recognize situations in which one quantity changes at a constant rate per unit interval relative to another.		4-1, 4-5
	HSF.LE.A.1.C Recognize situations in which a quantity grows or decays by a constant percent rate per unit interval relative to another.		4-1
HSF.LE.A.2	Construct linear and exponential functions, including arithmetic and geometric sequences, given a graph, a description of a relationship, or two input-output pairs (include reading these from a table).		1-5, 1-6, 1-7, 1-8, 1-9, 1-10, 4-1, 4-2, 4-3, 4-4, 4-6
HSF.LE.A.4	For exponential models, express as a logarithm the solution to $ab^{ct} = d$ where a, c, and d are numbers and the base b is 2, 10, or e; evaluate the logarithm using technology.		4-9, 4-10, 4-11, 4-14, 4-15, 4-16, 4-17, 4-18
HSF.LE.B Interpret expressions for functions in terms of the situation they model.			5-11
HSF.LE.B.5	Interpret the parameters in a linear or exponential function in terms of a context.		4-2, 4-7, 4-12, 4-13, 4-15
HSF.TF Trigonometric Functions			
HSF.TF.A Extend the domain of trigonometric functions using the unit circle.			6-3, 6-4, 6-5, 6-9
HSF.TF.A.1	Understand radian measure of an angle as the length of the arc on the unit circle subtended by the angle.		6-3, 6-4, 6-18
HSF.TF.A.2	Explain how the unit circle in the coordinate plane enables the extension of trigonometric functions to all real numbers, interpreted as radian measures of angles traversed counterclockwise around the unit circle.		6-5, 6-6, 6-10, 6-11, 6-12
HSF.TF.B Model periodic phenomena with trigonometric functions.			6-7, 6-14, 6-16, 6-19
HSF.TF.B.5	Choose trigonometric functions to model periodic phenomena with specified amplitude, frequency, and midline.*		6-13, 6-18, 6-19

Standards for Mathematical Content			Lesson(s)
HSF.TF.C Prove and apply trigonometric identities.			
HSF.TF.C.8		Prove the Pythagorean identity $\sin^2(\theta) + \cos^2(\theta) = 1$ and use it to find $\sin(\theta)$, $\cos(\theta)$, or $\tan(\theta)$ given $\sin(\theta)$, $\cos(\theta)$, or $\tan(\theta)$ and the quadrant of the angle.	6-5, 6-6
HSG Geometry			
HSG.GPE Expressing Geometric Properties with Equations			
HSG.GPE.B Use coordinates to prove simple geometric theorems algebraically.			
HSG.GPE.B.7		Use coordinates to compute perimeters of polygons and areas of triangles and rectangles, e.g., using the distance formula.*	7-6
HSS Statistics & Probability			
HSS.ID Interpreting Categorical & Quantitative Data			
HSS.ID.A Summarize, represent, and interpret data on a single count or measurement variable.			
HSS.ID.A.1		Represent data with plots on the real number line (dot plots, histograms, and box plots).	7-4, 7-5, 7-6
HSS.ID.A.2		Use statistics appropriate to the shape of the data distribution to compare center (median, mean) and spread (interquartile range, standard deviation) of two or more different data sets.	7-4, 7-5
HSS.ID.A.4		Use the mean and standard deviation of a data set to fit it to a normal distribution and to estimate population percentages. Recognize that there are data sets for which such a procedure is not appropriate. Use calculators, spreadsheets, and tables to estimate areas under the normal curve.	7-6, 7-7, 7-14, 7-15
HSS.ID.B Summarize, represent, and interpret data on two categorical and quantitative variables.			
HSS.ID.B.6		HSS.ID.B.6.A Fit a function to the data; use functions fitted to data to solve problems in the context of the data. Use given functions or choose a function suggested by the context. Emphasize linear, quadratic, and exponential models.	5-7, 5-8, 5-11
HSS.IC Making Inferences & Justifying Conclusions			
HSS.IC.A Understand and evaluate random processes underlying statistical experiments.			
HSS.IC.A.1		Understand statistics as a process for making inferences about population parameters based on a random sample from that population.	7-3
HSS.IC.A.2		Decide if a specified model is consistent with results from a given data-generating process, e.g., using simulation. For example, a model says a spinning coin falls heads up with probability 0.5. Would a result of 5 tails in a row cause you to question the model?	7-8
HSS.IC.B Make inferences and justify conclusions from sample surveys, experiments, and observational studies.			
HSS.IC.B.3		Recognize the purposes of and differences among sample surveys, experiments, and observational studies; explain how randomization relates to each.	7-1, 7-2, 7-3, 7-13
HSS.IC.B.4		Use data from a sample survey to estimate a population mean or proportion; develop a margin of error through the use of simulation models for random sampling.	7-9, 7-10, 7-11, 7-12
HSS.IC.B.5		Use data from a randomized experiment to compare two treatments; use simulations to decide if differences between parameters are significant.	7-13, 7-14, 7-15, 7-16
HSS.IC.B.6		Evaluate reports based on data.	7-1, 7-2, 7-13

Correlations to the Standards for Mathematical Practice, High School

Common Core State Standards for Mathematical Practice, High School
Correlated to Illustrative Mathematics, High School

Standards for Mathematical Practice	Illustrative Mathematics Lesson(s)			
MP1 Make sense of problems and persevere in solving them. Mathematically proficient students start by explaining to themselves the meaning of a problem and looking for entry points to its solution. They analyze givens, constraints, relationships, and goals. They make conjectures about the form and meaning of the solution and plan a solution pathway rather than simply jumping into a solution attempt. They consider analogous problems, and try special cases and simpler forms of the original problem in order to gain insight into its solution. They monitor and evaluate their progress and change course if necessary. Older students might, depending on the context of the problem, transform algebraic expressions or change the viewing window on their graphing calculator to get the information they need. Mathematically proficient students can explain correspondences between equations, verbal descriptions, tables, and graphs or draw diagrams of important features and relationships, graph data, and search for regularity or trends. Younger students might rely on using concrete objects or pictures to help conceptualize and solve a problem. Mathematically proficient students check their answers to problems using a different method, and they continually ask themselves, "Does this make sense?" They can understand the approaches of others to solving complex problems and identify correspondences between different approaches.	**Algebra 1** 1-2, 1-3, 1-12, 1-13, 1-14, 1-16, 2-3, 2-5, 2-11, 2-14, 2-19, 2-20, 2-24, 2-25, 2-26, 3-1, 3-2, 3-4, 3-10, 4-1, 4-6, 4-8, 4-11, 4-12, 4-16, 4-18, 5-2, 5-4, 5-5, 5-8, 5-9, 5-10, 5-13, 5-19, 5-21, 6-1, 6-2, 6-3, 6-4, 6-5, 6-14, 6-15, 6-17, 7-1, 7-2, 7-3, 7-9, 7-10, 7-21, 7-22, 7-24	**Algebra 1 Extra Support** 1-9, 1-10, 1-11, 1-12, 1-13, 1-15, 2-7, 2-15, 2-23, 3-1, 3-3, 4-3, 5-16, 6-6, 6-11, 7-1	**Geometry** 1-3, 1-4, 1-5, 1-8, 1-9, 1-10, 1-11, 1-12, 1-14, 1-18, 1-19, 1-22, 2-1, 2-4, 2-7, 2-8, 2-11, 2-12, 2-13, 2-15, 3-1, 3-5, 3-7, 3-11, 3-14, 3-15, 3-16, 4-4, 4-7, 4-10, 4-11, 5-2, 5-3, 5-8, 5-13, 5-15, 5-16, 5-18, 6-4, 6-6, 6-7, 6-12, 6-15, 7-2, 7-6, 7-9, 8-5	**Algebra 2** 1-1, 1-2, 1-5, 1-6, 1-7, 1-10, 1-11, 2-2, 2-4, 2-5, 2-8, 2-10, 2-12, 2-13, 2-15, 2-16, 2-17, 2-20, 2-22, 2-24, 2-25, 2-26, 3-3, 3-7, 3-9, 3-10, 3-12, 3-15, 3-19, 4-6, 4-7, 4-8, 4-18, 5-1, 5-4, 5-6, 5-8, 5-10, 5-11, 6-1, 6-4, 6-6, 6-12, 6-17, 6-19, 7-15
MP2 Reason abstractly and quantitatively. Mathematically proficient students make sense of quantities and their relationships in problem situations. They bring two complementary abilities to bear on problems involving quantitative relationships: the ability to decontextualize—to abstract a given situation and represent it symbolically and manipulate the representing symbols as if they have a life of their own, without necessarily attending to their referents—and the ability to contextualize, to pause as needed during the manipulation process in order to probe into the referents for the symbols involved. Quantitative reasoning entails habits of creating a coherent representation of the problem at hand; considering the units involved; attending to the meaning of quantities, not just how to compute them; and knowing and flexibly using different properties of operations and objects.	**Algebra 1** 1-1, 1-2, 1-3, 1-5, 1-11, 1-12, 1-13, 1-14, 1-15, 2-2, 2-3, 2-4, 2-6, 2-9, 2-10, 2-11, 2-12, 2-17, 2-18, 2-19, 2-20, 2-22, 2-23, 2-24, 2-26, 3-4, 3-5, 3-7, 3-8, 3-9, 4-2, 4-3, 4-4, 4-5, 4-6, 4-7, 4-8, 4-9, 4-10, 4-11, 4-15, 4-16, 5-3, 5-5, 5-6, 5-7, 5-8, 5-9, 5-10, 5-17, 5-18, 6-1, 6-3, 6-7, 6-8, 6-12, 6-14, 6-16, 7-1, 7-2, 7-3, 7-15, 7-16, 7-17, 7-18, 7-21	**Algebra 1 Extra Support** 1-1, 1-2, 1-3, 1-10, 1-12, 1-13, 1-14, 1-15, 1-16, 2-1, 2-2, 2-4, 2-6, 2-7, 2-9, 2-10, 2-12, 2-15, 2-18, 2-19, 2-20, 2-22, 2-23, 2-25, 3-3, 3-4, 3-6, 3-7, 3-8, 4-1, 4-2, 4-4, 4-6, 4-7, 4-8, 4-10, 4-11, 4-12, 4-16, 4-17, 5-1, 5-2, 5-5, 5-6, 5-8, 5-10, 5-12, 6-6, 6-7, 6-14, 7-1, 7-2, 7-4, 7-10, 7-17, 7-21, 7-24	**Geometry** 1-17, 2-10, 3-2, 3-6, 3-12, 3-13, 3-16, 4-3, 4-9, 5-1, 5-5, 5-7, 5-9, 5-12, 6-2, 6-8, 6-13, 7-3, 7-10, 7-11, 7-13, 8-5, 8-6, 8-7	**Algebra 2** 1-9, 1-10, 2-1, 2-3, 2-17, 2-18, 2-19, 2-20, 2-21, 3-1, 3-4, 3-7, 3-8, 4-1, 4-2, 4-3, 4-4, 4-7, 4-8, 4-13, 4-15, 4-16, 4-17, 5-5, 5-8, 6-1, 6-6, 6-8, 6-13, 6-14, 6-19, 7-1, 7-6, 7-10, 7-14, 7-15, 7-16

Standards for Mathematical Practice	Illustrative Mathematics Lesson(s)			
	Algebra 1	Algebra 1 Extra Support	Geometry	Algebra 2
MP3 Construct viable arguments and critique the reasoning of others. Mathematically proficient students understand and use stated assumptions, definitions, and previously established results in constructing arguments. They make conjectures and build a logical progression of statements to explore the truth of their conjectures. They are able to analyze situations by breaking them into cases, and can recognize and use counterexamples. They justify their conclusions, communicate them to others, and respond to the arguments of others. They reason inductively about data, making plausible arguments that take into account the context from which the data arose. Mathematically proficient students are also able to compare the effectiveness of two plausible arguments, distinguish correct logic or reasoning from that which is flawed, and—if there is a flaw in an argument—explain what it is. Elementary students can construct arguments using concrete referents such as objects, drawings, diagrams, and actions. Such arguments can make sense and be correct, even though they are not generalized or made formal until later grades. Later, students learn to determine domains to which an argument applies. Students at all grades can listen or read the arguments of others, decide whether they make sense, and ask useful questions to clarify or improve the arguments.	1-4, 1-11, 1-15, 2-3, 2-7, 2-11, 2-14, 2-15, 2-16, 2-23, 3-5, 3-6, 3-7, 4-3, 4-8, 5-11, 5-16, 5-18, 5-19, 6-2, 6-3, 6-12, 6-15, 6-16, 6-17, 7-5, 7-6, 7-13, 7-14, 7-18, 7-19, 7-20, 7-21, 7-23	1-3, 1-4, 1-5, 1-9, 1-10, 1-11, 1-12, 1-13, 1-15, 2-2, 2-6, 2-7, 2-11, 2-13, 2-15, 2-17, 2-22, 2-26, 3-1, 3-3, 3-5, 3-9, 3-10, 4-1, 4-8, 4-10, 4-16, 5-1, 5-3, 5-7, 5-8, 5-13, 5-15, 5-16, 5-17, 6-4, 6-12, 7-1, 7-5, 7-7, 7-9, 7-10, 7-11, 7-18, 7-19, 7-22	1-2, 1-4, 1-5, 1-10, 1-11, 1-12, 1-19, 1-20, 2-1, 2-3, 2-5, 2-7, 2-8, 2-9, 2-10, 2-13, 2-14, 3-4, 3-5, 3-8, 3-10, 3-14, 4-1, 4-8, 5-5, 5-6, 6-9, 6-10, 6-11, 6-14, 6-16, 7-6, 7-7, 8-2, 8-6, 8-11	1-1, 1-6, 2-4, 2-5, 2-9, 2-12, 3-8, 3-13, 3-14, 3-16, 3-18, 4-2, 4-4, 4-7, 4-14, 4-15, 5-1, 5-6, 5-11, 6-6, 6-19, 7-1, 7-2, 7-4, 7-8
MP4 Model with mathematics. Mathematically proficient students can apply the mathematics they know to solve problems arising in everyday life, society, and the workplace. In early grades, this might be as simple as writing an addition equation to describe a situation. In middle grades, a student might apply proportional reasoning to plan a school event or analyze a problem in the community. By high school, a student might use geometry to solve a design problem or use a function to describe how one quantity of interest depends on another. Mathematically proficient students who can apply what they know are comfortable making assumptions and approximations to simplify a complicated situation, realizing that these may need revision later. They are able to identify important quantities in a practical situation and map their relationships using such tools as diagrams, two-way tables, graphs, flowcharts and formulas. They can analyze those relationships mathematically to draw conclusions. They routinely interpret their mathematical results in the context of the situation and reflect on whether the results make sense, possibly improving the model if it has not served its purpose.	1-10, 1-14, 1-16, 2-1, 2-4, 2-9, 2-10, 2-12, 2-18, 2-20, 2-22, 2-23, 2-26, 3-4, 3-8, 3-9, 3-10, 4-1, 4-7, 4-8, 4-17, 4-18, 5-1, 5-4, 5-5, 5-8, 5-9, 5-11, 5-17, 5-21, 6-4, 6-6, 6-7, 6-11, 6-17, 7-1, 7-2, 7-17	1-9, 1-10, 1-11, 1-12, 1-13, 1-15, 2-7, 2-15, 3-1, 3-3, 4-9, 4-18, 5-9, 5-11, 5-19, 5-21, 6-14, 7-1, 7-6, 7-20	1-9, 2-12, 3-2, 3-16, 4-1, 5-8, 5-18, 7-14	1-9, 2-16, 2-22, 2-26, 4-15, 5-3, 5-8, 5-10, 5-11, 6-7, 6-19, 7-14, 7-15
MP5 Use appropriate tools strategically. Mathematically proficient students consider the available tools when solving a mathematical problem. These tools might include pencil and paper, concrete models, a ruler, a protractor, a calculator, a spreadsheet, a computer algebra system, a statistical package, or dynamic geometry software. Proficient students are sufficiently familiar with tools appropriate for their grade or course to make sound decisions about when each of these tools might be helpful, recognizing both the insight to be gained and their limitations. For example, mathematically proficient high school students analyze graphs of functions and solutions generated using a graphing calculator. They detect possible errors by strategically using estimation and other mathematical knowledge. When making mathematical models, they know that technology can enable them to visualize the results of varying assumptions, explore consequences, and compare predictions with data. Mathematically proficient students at various grade levels are able to identify relevant external mathematical resources, such as digital content located on a website, and use them to pose or solve problems. They are able to use technological tools to explore and deepen their understanding of concepts.	1-10, 1-16, 2-1, 2-9, 2-11, 2-12, 2-15, 2-17, 2-24, 2-26, 3-2, 3-3, 3-10, 4-13, 4-15, 4-17, 4-18, 5-1, 5-3, 5-5, 5-7, 5-8, 5-11, 5-18, 5-19, 5-21, 6-1, 6-2, 6-4, 6-6, 6-7, 6-12, 6-17, 7-3, 7-5, 7-17, 7-18, 7-20, 7-21	1-11, 1-14, 2-15, 2-24, 2-25, 5-5, 5-9, 5-12, 5-20, 6-4, 6-5, 6-14, 7-15, 7-17, 7-22, 7-24	1-2, 1-3, 1-4, 1-5, 1-6, 1-7, 1-9, 1-10, 1-11, 1-12, 1-13, 1-14, 1-17, 1-18, 1-19, 1-20, 1-21, 1-22, 2-1, 2-2, 2-4, 2-6, 2-7, 2-8, 2-9, 2-11, 2-13, 2-14, 2-15, 3-2, 3-7, 3-8, 3-9, 3-10, 3-11, 3-16, 4-1, 4-2, 4-3, 5-4, 5-15, 5-16, 6-1, 6-2, 6-6, 6-10, 6-11, 6-13, 6-14, 6-16, 6-17, 7-3, 7-5, 7-7, 7-14, 8-3, 8-6, 8-9, 8-10	1-4, 1-5, 1-6, 1-8, 1-10, 2-2, 2-8, 2-9, 2-20, 2-21, 2-26, 3-3, 3-9, 4-1, 4-5, 4-7, 4-8, 4-15, 4-16, 4-17, 5-6, 5-11, 6-7, 6-10, 6-11, 6-12, 6-14, 6-15, 6-18, 7-5, 7-6, 7-13, 7-14

Standards for Mathematical Practice	Illustrative Mathematics Lesson(s)			
MP6 Attend to precision. Mathematically proficient students try to communicate precisely to others. They try to use clear definitions in discussion with others and in their own reasoning. They state the meaning of the symbols they choose, including using the equal sign consistently and appropriately. They are careful about specifying units of measure, and labeling axes to clarify the correspondence with quantities in a problem. They calculate accurately and efficiently, express numerical answers with a degree of precision appropriate for the problem context. In the elementary grades, students give carefully formulated explanations to each other. By the time they reach high school they have learned to examine claims and make explicit use of definitions.	**Algebra 1** 1-1, 1-4, 1-11, 1-13, 1-15, 2-2, 2-5, 2-7, 2-12, 2-13, 2-17, 2-18, 2-21, 2-23, 2-25, 3-1, 3-4, 3-6, 3-7, 3-9, 4-1, 4-2, 4-3, 4-4, 4-6, 4-7, 4-8, 4-9, 4-11, 4-16, 5-2, 5-3, 5-9, 5-11, 5-13, 5-18, 5-21, 6-7, 6-9, 6-11, 6-14, 6-15, 6-16, 7-9, 7-10, 7-12, 7-13, 7-15, 7-16, 7-18, 7-20, 7-21, 7-22, 7-23	**Algebra 1 Extra Support** 1-4, 1-9, 1-12, 2-4, 2-6, 2-10, 2-11, 2-12, 2-16, 2-20, 2-24, 2-25, 3-6, 3-9, 3-10, 4-3, 4-5, 4-8, 4-9, 4-11, 5-8, 5-9, 5-11, 5-13, 5-21, 6-1, 6-6, 6-7, 6-12, 6-15, 6-17, 7-10	**Geometry** 1-1, 1-2, 1-6, 1-7, 1-9, 1-11, 1-13, 1-14, 1-16, 1-17, 1-18, 1-19, 1-20, 1-22, 2-2, 2-4, 2-5, 2-7, 2-8, 3-1, 3-3, 3-9, 3-10, 3-12, 3-15, 4-6, 4-7, 4-8, 5-1, 5-3, 5-6, 5-8, 5-12, 5-15, 5-17, 6-2, 6-5, 6-7, 6-8, 6-12, 6-13, 6-14, 7-1, 7-2, 7-3, 7-9, 7-12, 8-1, 8-5, 8-10	**Algebra 2** 1-2, 1-3, 1-7, 1-8, 1-9, 1-11, 2-1, 2-3, 2-6, 2-8, 2-9, 2-11, 2-14, 2-21, 2-22, 3-1, 3-3, 3-4, 3-5, 3-6, 3-8, 3-9, 3-11, 3-12, 3-14, 3-15, 3-18, 4-4, 4-6, 4-9, 4-10, 4-14, 4-15, 4-17, 5-1, 5-3, 5-5, 5-9, 6-1, 6-3, 6-8, 6-11, 6-13, 6-16, 6-17, 7-1, 7-2, 7-3, 7-4, 7-5, 7-15
MP7 Look for and make use of structure. Mathematically proficient students look closely to discern a pattern or structure. Young students, for example, might notice that three and seven more is the same amount as seven and three more, or they may sort a collection of shapes according to how many sides the shapes have. Later, students will see 7×8 equals the well-remembered $7 \times 5 + 7 \times 3$, in preparation for learning about the distributive property. In the expression $x^2 + 9x + 14$, older students can see the 14 as 2×7 and the 9 as $2 + 7$. They recognize the significance of an existing line in a geometric figure and can use the strategy of drawing an auxiliary line for solving problems. They also can step back for an overview and shift perspective. They can see complicated things, such as some algebraic expressions, as single objects or as being composed of several objects. For example, they can see $5 - 3(x - y)^2$ as 5 minus a positive number times a square and use that to realize that its value cannot be more than 5 for any real numbers x and y.	**Algebra 1** 1-2, 1-3, 1-10, 1-11, 1-13, 2-2, 2-3, 2-7, 2-11, 2-12, 2-13, 2-17, 2-20, 2-21, 2-23, 3-1, 3-3, 3-5, 3-6, 3-7, 3-8, 4-4, 4-8, 4-11, 4-12, 4-14, 4-16, 5-1, 5-2, 5-3, 5-6, 5-12, 5-13, 5-14, 5-18, 5-20, 6-1, 6-2, 6-3, 6-4, 6-5, 6-8, 6-9, 6-10, 6-12, 6-13, 6-15, 6-16, 6-17, 7-3, 7-4, 7-6, 7-7, 7-8, 7-9, 7-10, 7-11, 7-12, 7-13, 7-14, 7-19, 7-21, 7-22, 7-23	**Algebra 1 Extra Support** 1-2, 1-12, 2-2, 2-3, 2-5, 2-7, 2-8, 2-14, 2-16, 2-17, 2-20, 2-26, 3-1, 3-2, 3-7, 3-8, 4-12, 4-13, 4-14, 4-18, 5-1, 5-2, 5-4, 5-7, 5-10, 5-13, 5-14, 5-17, 5-18, 6-3, 6-4, 6-8, 6-10, 6-12, 6-15, 6-16, 7-3, 7-4, 7-6, 7-7, 7-8, 7-14, 7-15, 7-16, 7-22, 7-23	**Geometry** 1-2, 1-3, 1-5, 1-6, 1-7, 1-14, 1-15, 1-17, 1-19, 1-20, 1-21, 1-22, 2-2, 2-6, 2-9, 2-10, 2-11, 2-13, 2-14, 3-5, 3-6, 3-7, 3-9, 3-10, 3-11, 3-12, 3-14, 3-16, 4-5, 4-10, 5-6, 5-10, 5-11, 5-12, 5-13, 5-14, 5-15, 6-1, 6-2, 6-3, 6-5, 6-6, 6-8, 6-17, 7-5, 7-8, 7-10, 8-4, 8-5, 8-9	**Algebra 2** 1-2, 1-9, 1-11, 2-2, 2-3, 2-5, 2-6, 2-7, 2-8, 2-9, 2-10, 2-11, 2-13, 2-17, 2-20, 2-21, 2-25, 3-1, 3-4, 3-5, 3-6, 3-9, 3-11, 3-12, 3-13, 3-17, 3-18, 3-19, 4-2, 4-5, 4-6, 4-9, 4-10, 4-11, 4-14, 4-17, 4-18, 5-1, 5-2, 5-4, 5-5, 5-6, 5-7, 5-9, 5-11, 6-2, 6-5, 6-6, 6-7, 6-9, 6-11, 6-12, 6-15, 6-17, 6-18, 7-1, 7-7, 7-9, 7-11, 7-12, 7-13
MP8 Look for and express regularity in repeated reasoning. Mathematically proficient students notice if calculations are repeated, and look both for general methods and for shortcuts. Upper elementary students might notice when dividing 25 by 11 that they are repeating the same calculations over and over again, and conclude they have a repeating decimal. By paying attention to the calculation of slope as they repeatedly check whether points are on the line through (1, 2) with slope 3, middle school students might abstract the equation $(y - 2)/(x - 1) = 3$. Noticing the regularity in the way terms cancel when expanding $(x - 1)(x + 1)$, $(x - 1)(x^2 + x + 1)$, and $(x - 1)(x^3 + x^2 + x + 1)$ might lead them to the general formula for the sum of a geometric series. As they work to solve a problem, mathematically proficient students maintain oversight of the process, while attending to the details. They continually evaluate the reasonableness of their intermediate results.	**Algebra 1** 1-6, 1-9, 2-2, 2-3, 2-8, 2-11, 2-22, 4-15, 4-16, 5-3, 5-4, 5-12, 5-15, 5-16, 5-18, 5-20, 6-11, 6-2, 6-5, 6-6, 6-7, 6-11, 6-12, 6-13, 6-15, 7-4, 7-7, 7-8, 7-11, 7-14, 7-20	**Algebra 1 Extra Support** 2-2, 2-4, 2-8, 2-17, 2-21, 2-23, 4-15, 5-1, 5-3, 5-4, 5-17, 5-20, 6-1, 6-2, 6-5, 7-1, 7-4, 7-12, 7-13, 7-20	**Geometry** 1-6, 2-4, 2-7, 2-15, 3-6, 3-7, 3-10, 4-2, 4-11, 5-4, 6-4, 6-5, 7-4, 7-8, 8-8	**Algebra 2** 1-1, 1-5, 1-8, 2-15, 2-16, 2-19, 2-23, 3-1, 3-2, 3-11, 3-14, 4-1, 4-9, 4-10, 4-11, 4-12, 5-5, 5-10, 6-3, 6-4, 6-5, 6-10, 6-13, 6-16

Information for Families

The following information is available online, as a printable and customizable letter that teachers may choose to send home to families.

We'd like to introduce you to the *Illustrative Mathematics* curriculum. This problem-based curriculum makes rigorous high school mathematics accessible to all learners.

What is a problem-based curriculum?

In a problem-based curriculum, students spend most of their time in class working on carefully crafted and sequenced problems. Teachers help students understand the problems, ask questions to push their thinking, and orchestrate discussions to be sure that the mathematical takeaways are clear. Learners gain a rich and lasting understanding of mathematical concepts and procedures and experience applying this knowledge to new situations. Students frequently collaborate with their classmates—they talk about math, listen to each other's ideas, justify their thinking, and critique the reasoning of others. They gain experience communicating their ideas both verbally and in writing, developing skills that will serve them well throughout their lives.

This kind of instruction may look different from what you experienced in your own math education. Current research says that students need to be able to think flexibly in order to use mathematical skills in their lives (and also on the types of tests they will encounter throughout their schooling). Flexible thinking relies on understanding concepts and making connections between them. Over time, students gain the skills and the confidence to independently solve problems that they've never seen before.

What supports are in the materials to help my student succeed?

- Each lesson includes a lesson summary that describes the key mathematical work of the lesson and provides worked examples when relevant. Students can use this resource if they are absent from class, to check their understanding of the day's topics, and as a reference when they are working on practice problems or studying for an assessment.
- Each lesson is followed by a practice problem set. These problems help students synthesize their knowledge and build their skills. Some practice problems in each set relate to the content of the current lesson, while others revisit concepts from previous lessons and units. Distributed practice like this has been shown to be more effective at helping students retain information over time.
- Each lesson includes a few learning targets, which summarize the goals of the lesson. Each unit's complete set of learning targets is available on a single page, which can be used as a self-assessment tool as students progress through the course.
- Family support materials are included in each unit. These materials give an overview of the unit's math content and provide a problem to work on with your student.

What can my student do to be successful in this course?

Learning how to learn in a problem-based classroom can be a challenge for students at first. Over time, students gain independence as learners when they share their rough drafts of ideas, compare their existing ideas to new things they are learning, and revise their thinking. Many students and families tell us that while this was challenging at first, becoming more active learners in math helped them build skills to take responsibility for their learning in other settings.

Here are some ideas for encouraging your student:

- If you're not sure how to get started on a problem, that's okay! What can you try? Could you make a guess? Describe an answer that's definitely wrong? Draw a diagram or representation?
- If you're feeling stuck, write down what you notice and what you wonder, or a question you have, and then share that when it's time to work with others or discuss.
- Your job when working on problems in this class is to come up with rough-draft ideas and share them. You don't have to be right or confident at first, but sharing your thinking will help everyone learn. If that feels hard or scary, it's okay to say, "This is just a rough draft . . ." or "I'm not really sure but I think . . ."
- Whether you're feeling stuck or feeling confident with the material, listen to your classmates and ask them about their ideas. One way that learning happens is by comparing your ideas to other people's ideas, just like you learn about history by reading about the same events from different perspectives.
- At the end of class, or when you are studying, take time to write some notes for yourself. Ask yourself, "Do I understand the lesson summary? Do the learning targets describe me?" If not, write down a sentence like, "I understand up to . . . but I don't understand why . . ." Share it with a classmate, teacher, or other resource who can help you better understand.

We are excited to be able to support your student in their journey toward knowing, using, and enjoying mathematics. Feel free to contact me to arrange a time to discuss the specifics of your child's performance and how we can work together to help them succeed in this course.

Sincerely,

_____

(Teacher's Name)

_____

(Email/Phone)